PRACTICAL PROBLEMS in MATHEMATICS
for HEATING AND COOLING TECHNICIANS
Fifth Edition

Russell DeVore

DELMAR
CENGAGE Learning™

Australia • Brazil • Japan • Korea • Mexico • Singapore • Spain • United Kingdom • United States

DELMAR
CENGAGE Learning

Practical Problems in Mathematics for Heating and Cooling Technicians, Fifth Edition
Russell DeVore

Vice President, Career and Professional Editorial: **Dave Garza**

Director of Learning Solutions: **Sandy Clark**

Senior Acquisitions Editor: **James Devoe**

Managing Editor: **Larry Main**

Product Manager: **John Fisher**

Editorial Assistant: **Thomas Best**

Vice President, Career and Professional Marketing: **Jennifer McAvey**

Marketing Director: **Deborah S. Yarnell**

Marketing Manager: **Jimmy Stephens**

Marketing Coordinator: **Mark Pierro**

Production Director: **Wendy Troeger**

Production Manager: **Mark Bernard**

Content Project Manager: **David Plagenza**

Art Director: **Benj Gleeksman**

Technology Project Manager: **Christopher Catalina**

Production Technology Analyst: **Thomas Stover**

For product information and technology assistance, contact us at
Professional & Career Group Customer Support, 1-800-648-7450

For permission to use material from this text or product,
submit all requests online at **cengage.com/permissions**
Further permissions questions can be e-mailed to
permissionrequest@cengage.com

Library of Congress Control Number: 2008926960

ISBN-13: 978-1-4283-2428-2

ISBN-10: 1-4283-2428-3

Delmar
5 Maxwell Drive
Clifton Park, NY 12065-2919
USA

Cengage Learning products are represented in Canada by Nelson Education, Ltd.

For your lifelong learning solutions, visit **delmar.cengage.com**

Visit our corporate website at **cengage.com**

Notice to the Reader
Publisher does not warrant or guarantee any of the products described herein or perform any independent analysis in connection with any of the product information contained herein. Publisher does not assume, and expressly disclaims, any obligation to obtain and include information other than that provided to it by the manufacturer. The reader is expressly warned to consider and adopt all safety precautions that might be indicated by the activities described herein and to avoid all potential hazards. By following the instructions contained herein, the reader willingly assumes all risks in connection with such instructions. The publisher makes no representations or warranties of any kind, including but not limited to, the warranties of fitness for particular purpose or merchantability, nor are any such representations implied with respect to the material set forth herein, and the publisher takes no responsibility with respect to such material. The publisher shall not be liable for any special, consequential, or exemplary damages resulting, in whole or part, from the readers' use of, or reliance upon, this material.

Printed in Canada
2 3 4 X X 11 10 09

To
Carson Hall Sickieri, Makena Lynn Sickeri,
Haden Thomas DeVore, and Parker Player DeVore,
my legacies and my joys.

Contents

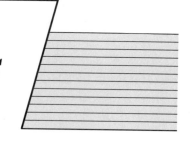

SECTION 11 GRAPHS / 249

SECTION 12 BILLS / 258

APPENDIX / 269

GLOSSARY / 290

ANSWERS TO ODD-NUMBERED PROBLEMS / 293

Preface

As a student I found that the easiest way to learn a subject was to relate it to something of interest to me. If I could see the relevance to a topic that I enjoyed, I could spend time on the subject without losing interest. In fact, many times, if a difficult subject was related to one that I like, it would seem easy to me. This is the philosophy behind the Practical Problems series. I have found that my students understand the concepts better if they can relate to the application of the topic.

I have tried hard to make each problem relevant to the heating and cooling field. You may not encounter each application in your experiences, but each of you will encounter many of these applications. It is my hope that this text will make the learning of math easier.

I would like to thank Steve Helba, Dawn Daugherty, and Christopher Chien for their assistance with this edition. I would also like to thank the people who reviewed the previous edition and made suggestions for improvement. Although all of the suggestions were not incorporated, they are appreciated and were seriously considered. In particular, I would like to thank the following instructors for their reviews and helpful suggestions:

Francois Nguyen, Saint Paul College, Saint Paul, MN

Carl Zitzer, Saint Paul College, Saint Paul, MN

Harvey Castelaz, MATC, Milwaukee, WI

Mark Smith, Lewis-Clark State College, Lewiston, ID

David Alvarez, Western Technical College, El Paso, TX

Carolyn Chapel, Western Technical College, LaCross, WI

ABOUT THE AUTHOR

RUSSELL DEVORE was chair of the Arts & Sciences Division at Trident Technical College in Charleston, South Carolina. He then taught in the Physics Department at Bloomsburg University in Pennsylvania. Dr. DeVore was named in the 1976 edition of *Who's Who in South Carolina* and *Outstanding Young Men of America*. He currently is a Shift Technical Advisor and Simulator Instructor at the training center of a nuclear power plant for PPL Susquehanna, LLC.

Introduction

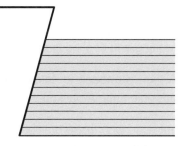

Before beginning this text, take some time to review the three topics discussed in this Introduction. This information may help you to make fewer errors when working math problems and may also make working the problems easier.

CALCULATORS

Today, the availability and low cost of calculators have made many math problems less of a chore. The modern calculators can do many calculations quickly. Thankfully, a complicated math calculation is no longer a big problem.

However, calculators can also have a "bad" side to them. People rely on them to do all of their math; they have gotten lazy when doing math. These people may get into trouble for a number of reasons.

First, when they are doing a math problem, these people have no idea if they have done it correctly. If an incorrect number was entered, or an incorrect button was pressed, these people would never know because they have no idea what kind of answer to expect. A couple of examples might clarify this point:

1. A person who relies totally on the calculator for math problems is multiplying 8.4 × 2.1. The calculator gives 4 as the answer. This person does not realize that a mistake was made by pressing the ÷ button instead of the × button, because he does not know that the correct answer should be about 16.

2. Someone doing a problem may not realize that a decimal point was not pressed or that one zero did not get entered into the calculator correctly when the number was being entered, because, once again, she has no idea what the correct answer should be. So, multiplying 8.4 × 2.1 may give an answer of 176.4 rather than 17.64 and the person doing the problem may not know that an error has been made.

So it is very important when working math problems with a calculator to have a rough idea what the answer should be.

There is a second drawback to calculators—they cannot work in fractions. If a person relies on the calculator all of the time, he will have no idea how to solve problems involving fractions because he has not practiced. Solving math problems is largely a matter of practicing; without practice, problems are hard to solve. Some people use the calculator to convert fractions to decimals; however, some fractions cannot be converted exactly. This means that the calculator answer will not be exactly correct. Since the heating and cooling field often requires linear measurements to be made, fractions will be encountered.

A third drawback to calculators is that they need power and they may run out of power at the worst possible time. A backup system should be ready, and that may be doing the math by hand.

Yet even with these drawbacks, calculators can be a big help in math. They can solve all problems with decimals and are really helpful when there are many numbers to add, subtract, or multiply. It is important to follow some basic rules:

1.   Get in the habit of thinking about what a reasonable answer would be. This makes it easier to notice when an error has been made.

2.   After entering the number into the calculator, check to see that you entered the correct number before pushing the operation button.

If you are going to buy a calculator, you will probably not need many of the special keys that some calculators have on them. The extra keys often make the calculators more expensive and may be a waste of money. More keys do not mean that the calculator is more accurate.

Each calculator comes with an instruction booklet. Use it. Read it completely. This is the only way that you will know exactly what your calculator will or will not do. One of the biggest shortcuts when working with most calculators is that when doing multiple operations, such as adding many numbers, you do not have to press the = button after each number is entered. Pressing the + button will cause the calculator to add the number just entered to the total already in the calculator and be ready to have another number entered. You have to press the = button only once, at the very end, to get the final sum. This same idea works for subtracting and multiplying. Remember that not all calculators work this way. Check the instruction booklet to make sure your calculator has this feature.

Calculators can be used for most of the problems in this book. When using the calculator, the answer should be estimated first. You should develop this habit of estimating the answer first and practice it whenever you do math. For practice, it is also recommended that every fifth problem be worked out by hand completely before using the calculator.

HINTS ABOUT ESTIMATING

As a check to see whether the correct number has been entered into the calculator and the correct button pushed, you should have an idea what the correct answer would be. You want to make this estimate as easy to determine as possible. What you really do is make a very easy math problem that is similar to the actual one you are trying to solve.

You want to make the problem so easy that you can do it in your head. To do this, you first need to "round" the numbers off. Usually, the better you round off, the closer you will get to the actual answer.

You want to round to one-digit numbers. So look at the next digit. If it is 5 or above, round your number up. If it is 4 or less, keep your number as it is. You want to round off 55 to 60 and 54 to 50. Round 149 to 100 because the second digit is a 4.

Next determine the mathematical operation that needs to be performed and perform it. You can "round off" the answer. This should be close to the actual answer. Remember that this is just the estimate. You have to go back and do the actual calculation. Examples of estimating answers are provided throughout the book.

INTRODUCTION TO MEASUREMENT

There are two parts to any measurement—the number (how many) and the unit (of what). We want to look at the units we use. Two major systems of measurement are used worldwide—the English and the metric. It is usual to work in one of the two systems at a time. You must be able to recognize the units and the system those units belong to. Let us look at some categories of measurement and what units belong to those categories.

One category has the same units for both measurement systems. That category is time. The units for time include seconds, minutes, hours, and days.

Mass or weight has English units of ounces, pounds, and tons. The metric units for mass or weight include grams and kilograms.

The English units for distance (length) include inches, feet, yards, and miles. The metric units for distance include centimeters, meters, and kilometers.

Area is the size of a flat surface. English units of area include square inches, square feet, and square yards. Metric units for area include square centimeters and square meters.

Volume is the size of the interior of a shape. English units of volume include cubic inches, cubic feet, and cubic yards. Metric units for volume include cubic centimeters and cubic meters.

Units can be a help in telling you if you are doing the problem correctly. If a problem asks how long it took to complete a job and you get an answer of $16.43, you know that you did something wrong, since dollars is not a measure of time (how long it took). So if a problem asks for the length of a piece of conduit and you determine the answer is 16 seconds, you know that you made an error. If you are asked to find the area of a window and you determine that it is 12 square feet, then there is a good chance that you worked the problem correctly.

We have to be careful with the units when working math problems. When adding or subtracting, the units must be the same so the answer will have the same unit. If they are not the same, an error will be made working the problem. On the other hand, when multiplying and dividing, the unit of the answer can be different from the units used in the calculation. As an example, a closet measures 3 feet by 4 feet. Notice that each of the measurements has the unit of feet. The floor area of this closet is 12 square feet.

Square feet is a different unit than feet, but in this case it is the correct unit. We found the area by multiplying the length by the width. In doing this, we also multiplied feet by feet and wound up with square feet. We still have to be careful that when we multiply similar quantities, we are using the same units. We should not multiply a length measured in feet by a width measured in inches. They should both be feet or both be inches.

There are times when two different quantities get multiplied or divided. The units are carried through with the numbers. As examples, torque is determined by multiplying a distance (measured in feet) by a force (measured in pounds). The result has the unit of foot-pounds. Speed is determined by dividing distance (measured in miles) by time (measured in hours). The result is speed with the unit of miles per hour. These are correct units.

When working problems throughout this text, pay attention to the units. Working the units properly is just as important as working the numbers correctly.

Whole Numbers

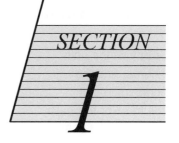

Unit 1 ADDITION OF WHOLE NUMBERS

BASIC PRINCIPLES OF ADDITION OF WHOLE NUMBERS

- Study addition of denominate numbers in Section I of the Appendix.

When a number is written, each numeral holds a certain place. Those places have names. For example, the number 4,567,893 has 3 in the units place, 9 in the tens place, 8 in the hundreds place, 7 in the thousands place, 6 in the ten thousands place, 5 in the hundred thousands place, and 4 in the millions place. Whole numbers are numbers that have nothing smaller than a unit; they have no fractional part to them.

Addition is the process of finding the total (sum) of two or more numbers.

When adding whole numbers by hand, it is best to place the numbers in columnar form. To do this, line the units places of each number underneath each other as shown below:

Example: Find the sum of the following:

$$
\begin{array}{r}
8 \\
27 \\
144 \\
6 \\
97 \\
\hline
\end{array}
$$

Let us solve this problem by hand and then estimate the answer and solve it using the calculator.

By lining up the units, all of the other places are also lined up.

Start with the units column. Add the numbers in that column. For this case we get a total of 32. Write the 2 in the units column of the answer.

$$
\begin{array}{r}
8 \\
27 \\
144 \\
6 \\
97 \\
\hline
2
\end{array}
$$

If the total is greater than ten, as in this case, carry the tens number to the top of the next column and then add that number along with the other numbers in that column. In this case we get 18. Put the 8 in the tens column of the answer and carry the 1 to the hundreds column.

$$
\begin{array}{r}
^{3}8 \\
27 \\
144 \\
6 \\
97 \\
\hline
82
\end{array}
$$

Do the same for each column to get the final answer.

$$
\begin{array}{r}
^{1\ 3}8 \\
27 \\
144 \\
6 \\
97 \\
\hline
282
\end{array}
$$

To estimate the answer for this problem, rewrite the problem as

$$
\begin{array}{r}
8 \\
27 \\
144 \\
6 \\
97 \\
\hline
\end{array}
\quad \Rightarrow \quad
\begin{array}{r}
10 \\
30 \\
100 \\
10 \\
100 \\
\hline
\end{array}
$$

Adding this second problem, one easily gets 250. So the answer to the actual problem is going to be near 250. (We now know 2,500 is too large and 25 is too small, since the answer should be close to 250. If the calculator gets 644 as the answer, an error was made, since 644 is not close to 250.)

Adding the **original** numbers on the calculator gives us 282. This is close to 250, so the answer is correct.

PRACTICAL PROBLEMS

Note: In problems 1–12, add the numbers.

1.   625
     +351

2.   2,273
      5,402
    + 313

3.   1,302
      6,001
        264
    +1,412

4.   129 sq in
       453 sq in
     +287 sq in

5.   567 meters
       180 meters
        35 meters
     +208 meters

6.   3,413 gallons
      9,020 gallons
        787 gallons
      4,121 gallons
    +2,568 gallons

7.   247 + 501 + 131   _____

8.   632 + 51 + 3   _____

9.   7,204 + 60 + 13 + 21   _____

10.   656 grams + 804 grams + 222 grams   _____

11.   1,717 feet + 485 feet + 9,204 feet + 339 feet   _____

12.   13 minutes + 491 minutes + 877 minutes   _____

13.   This house is heated by electric baseboard heaters. Find in feet the total length of heaters needed.   _____

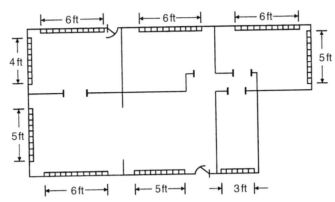

14.   A forced air system takes return air at 62 degrees Fahrenheit (°F). It heats the air 59°F. What is the temperature of the heated air blown out to the room?   _____

15.   Find in feet the total distance air travels through this duct.   _____

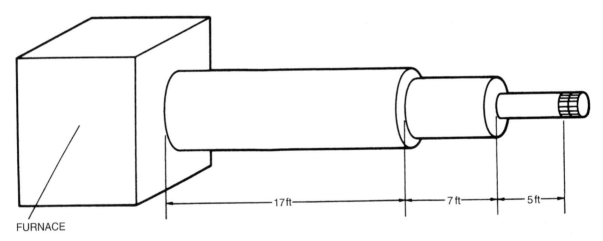

FURNACE

16.   Pipe lengths of 2 inches, 8 inches, 108 inches, 8 inches, and 24 inches are used to make a conduit for wires to a heat pump. How many inches of pipe are needed for this conduit?   _____

17. The condensing unit of an air conditioner is wired to the power source. The wires run along the wall of the house. Find in feet the amount of wire that is used. _____

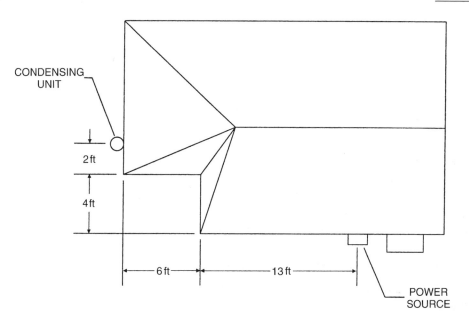

18. In one month, a technician works 43 hours, 41 hours, 40 hours, and 41 hours. What is the total number of hours the technician works? _____

19. Find in cubic feet (cu ft) the total volume of this house. _____

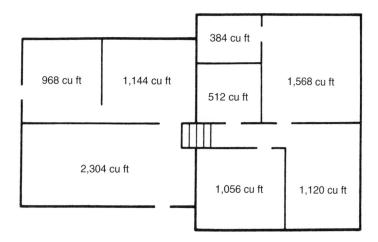

20. Starting from the shop, a technician makes five stops in one day and returns to the shop. Find in miles the total distance the technician travels. _____

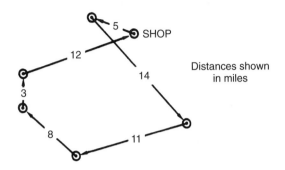

SHOP

Distances shown in miles

21. A forced convection heating system is installed in a house. The lengths of 6-inch circular duct that are needed are: kitchen, 6 feet; dining room, 12 feet; living room, 3 feet; master bedroom, 5 feet; second bedroom, 7 feet; third bedroom, 13 feet. How many feet of 6-inch round duct are needed? _____

22. A repair truck carries 121 feet of $\frac{3}{4}$-inch copper tubing, 43 feet of $\frac{1}{8}$-inch copper tubing, 76 feet of 1-inch conduit pipe, and 112 feet of $\frac{1}{2}$-inch polyvinylchloride (PVC) pipe. What is the total length of material on the truck? _____

23. An air-conditioning shop orders the following amounts of refrigerant: 125 pounds of R-134a, 150 pounds of R-125, 70 pounds of R-124, 90 pounds of R-32, and 120 pounds of R-152a. What is the total weight of refrigerant ordered? _____

24. Truck 1 carries seven 1-inch check valves, truck 2 carries five 1-inch check valves, and truck 3 carries eight 1-inch check valves. The warehouse has nineteen 1-inch check valves. What is the total inventory for 1-inch check valves? _____

25. Last year, the Keep Kool Company's five repair trucks covered 7,252 miles; 8,917 miles; 4,266 miles; 7,793 miles; and 8,214 miles. What is the total mileage the Keep Kool Company should report for last year? _____

Unit 2 SUBTRACTION OF WHOLE NUMBERS

BASIC PRINCIPLES OF SUBTRACTION OF WHOLE NUMBERS

- Study subtraction of denominate numbers in Section I of the Appendix.

Subtraction is the process of finding the difference between two numbers.

When subtracting, place the smaller number underneath the larger one, and line up the units columns. The other columns will also be aligned. Beginning with the units column, subtract the lower number from the upper one.

Example: Find the difference between 1,749 and 563.

$$
\begin{array}{r}
1,749 \\
-\ \ \ 563 \\
\hline
6
\end{array}
$$

It is not possible to subtract 6 from 4, so "borrow" 1 from the 7, making the 7 a 6. The 1 becomes a 10 and is added to the 4, making the 4 a 14. Now 6 can be subtracted from 14. The 5 gets subtracted from 6.

7 becomes 6
4 becomes 14
$$
\begin{array}{r}
1,749 \\
-\ \ \ 563 \\
\hline
1,186
\end{array}
$$

We could estimate the answer to this problem either of two ways. Remember that estimating is to make a problem that you can do easily.

First way: Change

$$
\begin{array}{r}
1,749 \\
-\ \ \ 563
\end{array}
\quad \text{to} \quad
\begin{array}{r}
2,000 \\
-\ \ \ 600
\end{array}
$$

Subtracting 6 from 20 gives 14, so the estimated answer is 1,400.

Second way: Change

$$
\begin{array}{r}
1,749 \\
-\ \ \ 563
\end{array}
\quad \text{to} \quad
\begin{array}{r}
1,700 \\
-\ \ \ 600
\end{array}
$$

Subtracting 6 from 17 gives 11, so the estimated answer is 1,100.

Either way is acceptable. They both have an answer larger than 1,000 but smaller than 1,500.

Using the calculator, we get 1,749 − 563 = 1,186. This is close to the answer that was estimated.

PRACTICAL PROBLEMS

Note: In problems 1–12, subtract the numbers.

1.   87
    −23

2.   557
    −125

3.   927
    −216

4.   481 kilometers
    −328 kilometers

5.   1,735 quarts
    − 466 quarts

6.   904 yards
    −149 yards

7.   853 − 522 _____

8.   489 − 169 _____

9.   796 − 81 _____

10.   526 cu ft − 440 cu ft _____

11.   748 sq cm − 159 sq cm _____

12.   1,104 seconds − 378 seconds _____

13.   This indoor-outdoor thermometer shows the temperature readings for a winter day. What is the temperature difference? _____

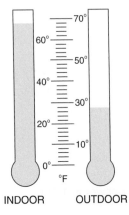

INDOOR   OUTDOOR

14. It takes 104 hours to install a heating system in an office building. The installers have already worked 36 hours. How many more hours are needed to finish the installation?

15. A cylinder containing refrigerant R-134a weighs 22 pounds. When empty, the cylinder weighs 4 pounds. How many pounds of R-134a are in the cylinder?

16. This room needs 19 feet of electric baseboard heaters. Lengths of 6 feet and 7 feet are already placed. What length of baseboard heater still must be placed?

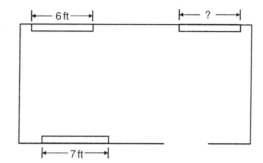

17. Air flows through the main duct of this system at a rate of 220 cubic feet per minute (cu ft/min). In one of the branches, the rate is 96 cu ft/min. Find in cubic feet per minute the rate that the air flows through the other branch.

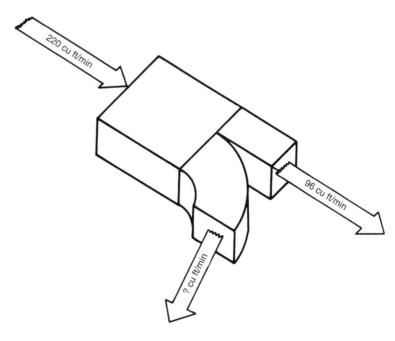

18.   A heating and air-conditioning repair shop keeps 1,500 feet of ¾-inch copper tubing in stock. An inventory shows that there are only 623 feet on hand. How many feet of tubing need to be ordered?   _____

19.   A 162-foot roll of *#10* wire is used when installing a residential air-conditioning unit. Lengths of 17 feet and 18 feet are cut from the roll. How many feet are left?   _____

20.   This condensing coil has 3 grill openings. The total length of the openings is 57 inches. Find the length of the center opening.   _____

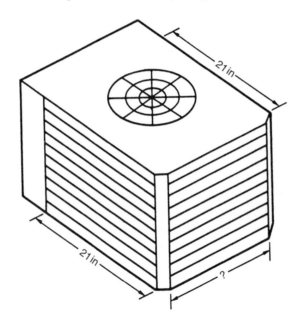

21.   A full oil tank holds 280 gallons of *#2* fuel oil. During 4 months, the amounts of fuel oil used are: 0 gallons, 18 gallons, 53 gallons, and 123 gallons. How much fuel oil is left in the tank?   _____

22.   An installer has 84 strap hangers in the truck. On a job, 32 straps are used on hot water supply lines and 29 on return lines. The installer also uses 15 straps on an electrical conduit that carries wires to the furnace. Find the number of straps that are left in the truck.   _____

23.   A large house has 4,676 sq ft of floor space and is divided into 3 heating/cooling zones. The first zone, the upstairs, has 1,625 sq ft. The main living area is the second zone covering 2,130 sq ft of space. The third zone is the basement where the rec room, laundry, half bath, and home shop are located. How much floor space is included in the third zone?   _____

24. An air-conditioning installer orders 150 sheets of 3-foot by 8-foot sheet metal. The installer keeps this record:

NUMBER OF SHEETS	USE
13	6-inch circular ducts
19	4-inch circular ducts
25	10-inch × 12-inch rectangular ducts
8	3-inch × 5-inch rectangular ducts

How many sheets of metal are left?   _____

25. A main air supply plenum for a room supplies air to 4 diffusers. The plenum has a flow of 150 cu ft/min. Diffuser 1 has a flow of 45 cu ft/min, diffuser 2 has a flow of 40 cu ft/min, and diffuser 3 has a flow of 25 cu ft/min. What is the airflow through the fourth diffuser?   _____

Unit 3 MULTIPLICATION OF WHOLE NUMBERS

BASIC PRINCIPLES OF MULTIPLICATION OF WHOLE NUMBERS

- Study multiplication of denominate numbers in Section I of the Appendix.

When multiplying two numbers, place one under the other and line up their units columns.

Example 1: An office building has 47 offices in it. Each identical office has a total of 651 sq ft of window glass in it. What is the total amount of window glass in the office building?

This is a multiplication problem. We multiply the amount of window glass in each office by the number of offices there are in the building. Multiply 651 sq ft by 47.

$$\begin{array}{r} 651 \text{ sq ft} \\ \times\ 47 \\ \hline \end{array}$$

Start with the unit number of the second number and multiply it by the entire first number. Carry the number for any products larger than ten.

$$\begin{array}{r} \overset{3}{}651 \\ \times\ 47 \\ \hline 57 \end{array} \quad \text{(multiply 7 times 6 and then add 3)} \quad \begin{array}{r} \overset{3}{}651 \\ \times\ 47 \\ \hline 4557 \end{array}$$

Then multiply the tens number of the second number by the entire first number. Line the first number that you write down directly under the same column that you are multiplying by; in this case, the number would be directly under the tens column.

$$\begin{array}{r} \overset{2}{}651 \\ \times\ 47 \\ \hline 4557 \\ 2604 \\ \hline \end{array}$$

Finally, add the numbers that you have written down, column by column. This will give you the final answer. (Notice that the answer is written with commas at the proper places. This should be done while the answer is being written.) Do not forget to include the measurement units in the answer.

$$
\begin{array}{r}
651 \text{ sq ft} \\
\times\ 47 \\
\hline
4\ 557 \\
26\ 04 \\
\hline
30{,}597 \text{ sq ft}
\end{array}
$$

To estimate an answer for this problem, change

$$
\begin{array}{r}
651 \text{ sq ft} \\
\times\ 47 \\
\hline
\end{array}
\qquad
\begin{array}{r}
700 \text{ sq ft} \\
\times\ 50 \\
\hline
35{,}000 \text{ sq ft}
\end{array}
$$

We just have to multiply 5 × 7, giving 35, so the correct answer is around 35,000 sq ft. Using the calculator we get 30,597 sq ft.

If three numbers are to be multiplied, such as 2 × 3 × 4, multiply the first two numbers together. Then multiply that answer by the third number.

Example 2: On an installation job, 2 men each worked 4 hours each day on 3 different days. What is the total number of hours worked on the job?

The answer is found by multiplying 2 × 4 hours × 3. The answer will have the unit of hours.

$$
\begin{array}{r}
2 \\
\times 4 \text{ hours} \\
\hline
8 \text{ hours}
\end{array}
\qquad
\begin{array}{r}
8 \text{ hours} \\
\times\ 3 \\
\hline
24 \text{ hours}
\end{array}
$$

With numbers this small, it is as easy to do the math in your head as to estimate it and then use the calculator. It is also easy to do this on a calculator. You just press 2 × 4 × 3 = and get 24.

PRACTICAL PROBLEMS

> **Note:** In problems 1–12, multiply the numbers.

1. $\begin{array}{r} 73 \\ \times\ 3 \\ \hline \end{array}$

2. $\begin{array}{r} 41 \\ \times 29 \\ \hline \end{array}$

3. $\begin{array}{r} 662 \\ \times 706 \\ \hline \end{array}$

4. $\begin{array}{r} 56 \text{ inches} \\ \times\ 9 \\ \hline \end{array}$

5. $\begin{array}{r} 359 \text{ millimeters} \\ \times\ 48 \\ \hline \end{array}$

6. $\begin{array}{r} 499 \text{ gallons} \\ \times 376 \\ \hline \end{array}$

7. 19 × 65   _____

8. 478 × 41   _____

9. 509 × 92   _____

10. 840 liters × 760   _____

11. 321 sq ft × 251 × 6   _____

12. 420 × 537 × 86 hours   _____

13. There are 144 electrical connectors in a box. How many connectors are there
    in 7 boxes?   _____

14. An apartment building has 8 apartments in it. Each apartment has 7 duct
    openings that need diffusers. How many diffusers are needed for the entire
    building?   _____

15. A technician charges $16 per hour for labor. How much should be charged for
    a 17-hour job?   _____

16. A can of refrigerant 134a (R-134a) for car air conditioners contains 14 ounces.
    A case of 24 cans has a total of how many ounces of R-134a?   _____

17. There are 36 cylinders of refrigerant R-125 in a stockroom. Each cylinder
    contains 137 pounds of refrigerant. How many pounds of R-125 are in the
    stockroom?   _____

18. Using one 3-foot by 8-foot piece of sheet metal, an installer makes four 6-inch
    circular ducts. Each duct is 3 feet long. How many ducts can be made from
    26 sheets of metal?   _____

19. A shopping mall has 26 air-conditioning units. Each unit can remove 37,000
    British thermal units (Btu) of heat each hour. Find in Btu the total amount of
    heat that can be removed each hour.   _____

20. Burning 1 gallon of #1 fuel oil produces 137,000 Btu of heat. How many
    British thermal units of heat are produced when 250 gallons are burned?   _____

21. A ground source heat pump system uses a horizontal ground loop consisting
    of 4 trenches for its heat source. Each trench is 95 feet long. What is the total
    length of trench for this system?   _____

22. *Latent heat of fusion* is the heat given up as a substance freezes. The latent heat of fusion for water is 144 British thermal units per pound (Btu/lb). This means that 144 Btu are given up for each pound of water that freezes. How many British thermal units of heat are given up when 4,000 pounds of water freeze?

_____

23. In an air-conditioning repair shop, more storage space is needed. To make space, 7 units of shelves are purchased. Each 5-shelf unit is 6 feet long. How many feet of shelf space are added?

_____

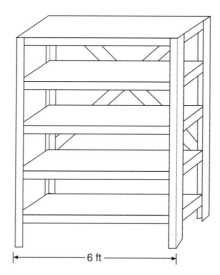

|← ——— 6 ft ——— →|

24. An office building is being built. One main supply duct will supply the ventilation flow for the identical offices shown. Each office flow is supposed to be 156 cu ft/min. How many cu ft/min should the main supply plenum be designed to handle?

_____

Office	Office	Office	Office
Hall				

Office	Office	Office	Office
Office	Office	Office	Office
Hall				

25. One man-hour is one man working for 1 hour. A housing development has 14 buildings. Each building has 6 condominiums in it. Each condominium needs a heating/air-conditioning system installed in it. Each system installation will take 23 man-hours to complete. The contractor must plan for how many man-hours to complete the job?   _____

Unit 4 *DIVISION OF WHOLE NUMBERS*

BASIC PRINCIPLES OF DIVISION OF WHOLE NUMBERS

- Study division of denominate numbers in Section I of the Appendix.

Division is probably the most difficult of the mathematical operations. But it can be done correctly time after time if care is taken to do the correct procedure each time. Each of the numbers in a division problem has a special name:

$$\text{DIVISOR })\overline{\text{DIVIDEND}}^{\text{QUOTIENT}}$$

The problem could also have been written as DIVIDEND ÷ DIVISOR = QUOTIENT.

To perform division, first write the problem down.

Example: Divide 194,256 by 57.

$$57 \,)\overline{194,256}$$

Divide the divisor into the first number of the dividend. How many 57s are in 1? In this case, the dividend is too small. The divisor will not go into the number. So try the divisor into the first two numbers of the dividend. How many 57s are in 19? Again the 19 is too small. Next try 57 into 194. There are three 57s in 194. Place the 3 in the quotient directly above the 4 in the dividend. There will then be a number in the quotient for each additional number in the dividend.

$$57 \,)\overline{194,256}^{\;\;\;3}$$

Now multiply 57 by 3 and place this number under 194. Subtract this number from 194.

$$
\begin{array}{r}
3 \\
57 \,)\overline{194,256} \\
\underline{171} \\
23
\end{array}
$$

Bring down the 2 from the dividend to make 232. Divide 57 into this number. This gives 4. Multiply 57 by 4 and subtract this number from 232.

$$
\begin{array}{r}
3\,4 \\
57 \,)\overline{194,256} \\
\underline{171} \\
23\,2 \\
\underline{22\,8} \\
4
\end{array}
$$

Bringing down the next number makes 45. Dividing 57 into 45 gives nothing; 45 is too small. However, instead of just bringing down the next number like we did at the beginning of the problem, a 0 is placed in the quotient above the 5. (There must be a number in the quotient for each number in the dividend after the first number has been placed in the quotient.) After this has been done, bring down the next number and continue.

$$
\begin{array}{r}
3{,}408 \\
57\,\overline{)194{,}256} \\
171\phantom{{,}256} \\
\hline
23\,2 \\
22\,8 \\
\hline
456 \\
456 \\
\hline
0
\end{array}
$$

To estimate the answer, change
$$57\,\overline{)194{,}256} \quad \text{to} \quad 60\,\overline{)200{,}000}$$

Now, just like the longhand part of the real problem, 60 does not go into 2, and 60 does not go into 20, but 60 does go into 200 three times leaving 20. The 3 is placed over the last 0 in 200. The next 0 is brought down which makes it 60 into 200 again, so we get another 3 with 20 left over. Continuing this gives an estimated quotient of 3,333. Notice that this is not an exact number. There is still a remainder. But we do not have to worry about that, since we are only getting an estimated answer.

Using the calculator, 194,256 ÷ 57 = 3,408. This is close to our estimated answer.

PRACTICAL PROBLEMS

Note: problems 1–12, divide the numbers.

1. $3\,\overline{)69}$

2. $13\,\overline{)442}$

3. $171\,\overline{)4{,}446}$

4. $\overline{)426}$ cu cm

5. $42\,\overline{)14{,}952}$ Btu

6. $183\,\overline{)89{,}304}$ amperes

7.   86 ÷ 2   _____

8.   918 ÷ 27   _____

9.   21,239 ÷ 317   _____

10.   392 days ÷ 7   _____

11.   2,268 pounds ÷ 63   _____

12.   245,971 cu in ÷ 649   _____

13.   A crate contains 4 compressors. The 4 compressors weigh 332 pounds. What
      is the weight of one compressor?   _____

14.   An apartment complex with 224 apartments is being built. Each workday, a
      heating and air-conditioning system can be installed in 7 apartments. How
      many workdays will it take to install all of the systems?   _____

15.   A fuel oil tank, when filled, weighs 1,900 pounds. It stands on 4 legs as
      shown. What weight must the floor support under each leg; that is, what is the
      weight each leg supports?   _____

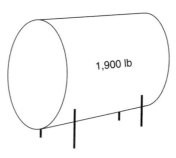

16.   New tires are purchased for service trucks. Each service truck is given 4 tires
      and a spare. If there are 65 new tires, how many trucks will receive new tires?   _____

17.   This 1,800-sq ft attic is to be insulated. One roll of 6-inch thick insulation
      covers 40 sq ft of the attic. How many rolls are needed to insulate the entire
      attic?   _____

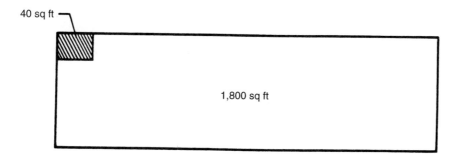

18. The weight of 75 pieces of 3-foot by 8-foot sheet metal is 600 pounds. Find the weight of one sheet.

_____

19. To complete the installation of a heating system, 728 hours are needed. Each of the 7 installers works 8 hours a day. How many days will it take to complete the installation?

_____

20. One hanger strap is used for the beginning of this duct. The other hanger straps are placed 5 feet apart. How many hanger straps are needed?

_____

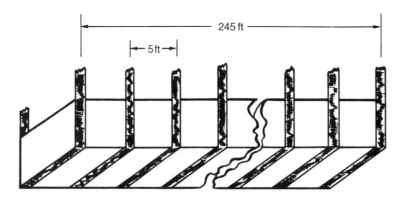

21. A 3-foot by 8-foot piece of sheet metal is used to make 4-inch circular ducts. Each duct is 8 feet long. How many sheets are needed to make 224 feet of the 4-inch circular ducts?

_____

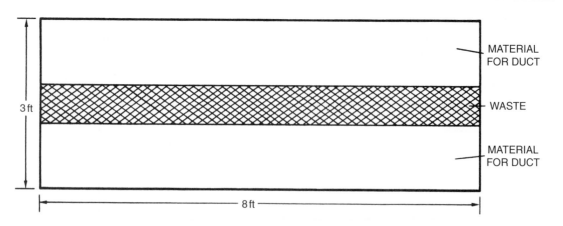

22. The Energy Efficiency Rating (EER) of an air-conditioning unit is defined as the ratio of the heat removed in Btu/hr divided by the energy used in watts. A window air conditioner is rated at 5,170 Btu. It uses 470 watts of power. What is this unit's EER?

_____

23. When 1 cu ft of natural gas is burned, 1,050 Btu of heat are produced. In one day, a building uses 761,250 Btu of heat. How many cubic feet of gas are burned?   _____

24. Air flows into this room at the rate of 71 cu ft/min. How many minutes will it take to change the air in this room?   _____

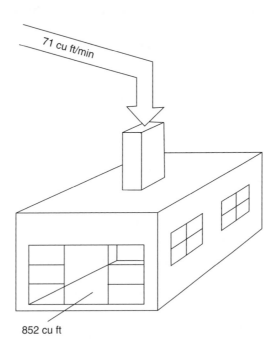

852 cu ft

25. A cylinder of R-134a contains 276 ounces of refrigerant. The cylinder is used to recharge car air conditioners. Each recharge takes 12 ounces. How many car air conditioners can be recharged from this cylinder?   _____

Unit 5 COMBINED OPERATIONS WITH WHOLE NUMBERS

BASIC PRINCIPLES OF COMBINED OPERATIONS WITH WHOLE NUMBERS

- Review and apply the principles of addition, subtraction, multiplication, and division of whole numbers to the problems.

The word problems in this unit are more like real-life problems than those in the preceding units. That is because the first thing you have to do when given a problem is to decide what you need to do to solve the problem. In this unit, that means you have to decide whether to add, subtract, multiply, or divide. Then you have to decide what numbers you are going to add, subtract, multiply, or divide—and then you perform the math operation. You also have to ask yourself, "Are there units that I need to put with the answer?"

Here are some hints that may make the decision about what operation to perform a little easier. These hints will not work in every case, but they may help.

1. If the problem asks you to find the **difference**, it is a subtraction problem.

2. If the problem asks for the **total**, it is either an addition problem or a multiplication problem.

You should develop a series of hints that work for you. You do this by practicing and doing lots of problems.

Example: On Monday, the storeroom had 54 thermostats in stock. Putting a heating system in a new house on Tuesday required 9 thermostats. Repairing failed thermostats on a job on Wednesday used 4 thermostats. On Thursday, 6 thermostats were drawn to restock the truck, so the storeroom ordered more. On Friday, three boxes of 8 thermostats each arrived in the storeroom. At the end of Friday, how many thermostats does the storeroom have?

To solve this problem, multiple operations are needed. Since the 9 thermostats, the 4 thermostats, and the 6 thermostats were removed from the storeroom, those numbers should be subtracted from the 54 thermostats that were in stock. Then on Friday, three boxes each having 8 thermostats arrived in the storeroom. These have to be added to the ones that are there.

In estimating the answer, 9 and 4 and 6 all need to be subtracted from 54. Rounding the numbers makes the estimate $50 - 10 - 0 - 10 = 30$. Next, $3 \times 8 = 24$ is rounded off to 20.

$$30 + 20 = 50$$

So the answer should be around 50 thermostats.

Doing the problem by hand or using a calculator:

$$54 - 9 - 4 - 6 = 35$$
$$3 \times 8 = 24$$
$$35 + 24 = 59$$

There are 59 thermostats in the storeroom at the end of Friday.

PRACTICAL PROBLEMS

1.
```
  635
  211
  583
+ 444
```

2.
```
  4,812
  9,303
    295
+ 6,450
```

3.
```
  643
- 218
```

4.
```
  833
- 394
```

5.
```
  7,035
×   476
```

6.
```
  2,884
× 3,526
```

7. $39 \overline{)936}$

8. $252 \overline{)124,992}$

9. 137 inches + 5,048 inches + 752 inches   _____

10. 265 hours − 88 hours   _____

11. 431 seconds × 726 × 89   _____

12. 7,781 kilometers ÷ 31   _____

13.   An addition is added to a room. The addition contains 758 cu ft of space.
What is the new volume of the room in cubic feet?   _____

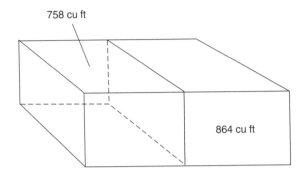

758 cu ft

864 cu ft

14.   The heat loss, in Btu/hr, for two different types of walls is given. Find in Btu/hr
the difference in the heat losses.   _____

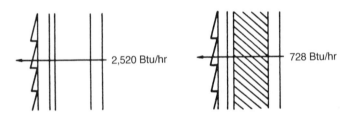

2,520 Btu/hr

728 Btu/hr

15.   A stockroom has 19 boxes of valves. Each box has 4 shut-off valves. What is
the total number of shut-off valves in the stockroom?   _____

16.   A family keeps this record of the number of gallons of fuel oil used to heat
their home:

MONTH	GALLONS USED
September	35
October	82
November	144
December	193

Find the total number of gallons used.   _____

17. Air flows into a room at the rate of 80 cu ft/min. The volume of the room is 960 cu ft.

a. How many minutes does it take to change the air in the room?   _____

b. How many times will the air be changed in 60 minutes?   _____

18. The storage capacity of a refrigerator is found by using the inside dimensions of the refrigerator.

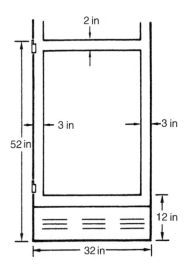

a. Find in inches the inside width of this refrigerator.   _____

b. Find in inches the inside height of this refrigerator.   _____

19. At high speed, a blower delivers 2,580 cu ft/min. This volume is divided equally among 12 ducts. Find in cubic feet the amount of air that flows through each duct every minute.   _____

20. A hot water baseboard radiator has 12 fins per inch. How many fins are in 72 inches of the radiator?   _____

21. An office building with three air-conditioning units has an air-conditioning load of 121,000 Btu/hr. The first unit handles 41,000 Btu/hr, and the second handles 42,000 Btu/hr. How many Btu/hr does the third unit handle?   _____

22.   A forced air heating system is used in this mobile home. Find in cu ft/min the airflow that the central blower must supply.   _____

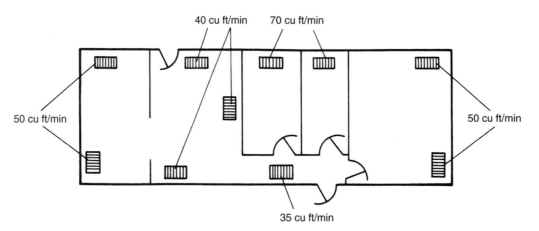

40 cu ft/min   70 cu ft/min

50 cu ft/min   50 cu ft/min

35 cu ft/min

23.   An installer finds that it takes 5 minutes to cut out a duct. It then takes 4 minutes to bend the metal and 7 minutes to lap and seal it. If the installer works 50 minutes out of each hour, how many ducts can be made during an 8-hour shift?   _____

24.   A ground source heat pump system uses wells as its heat source. Four wells are needed for the system. They are placed in a straight line going out from the house. The first well is 11 feet from the house. The wells are 18 feet apart. How far from the house is the farthest well?   _____

25.   A 1-ton air conditioner is a unit that will remove the amount of heat needed to melt one ton of ice in 24 hours. A ton has 2,000 pounds in it.

    a. If 144 Btu of heat are needed to melt 1 pound of ice, how many Btus would a 1-ton air conditioner remove in 24 hours?   _____

    b. How many Btus would a 1-ton air conditioner remove in 1 hour?   _____

Common Fractions

Unit 6 ADDITION OF COMMON FRACTIONS

BASIC PRINCIPLES OF ADDITION OF COMMON FRACTIONS

- Study addition of denominate numbers in Section I of the Appendix.

Fraction problems are best worked longhand. Some fractions do have exact decimal equivalents and, therefore, can be converted to numbers that can be worked with on the calculator. Not all fractions are like that, so it is best to learn how to solve problems involving fractions without the use of the calculator.

We will look at how to solve some fraction problems using the calculator in Unit 15. Before that, let us solve fraction problems longhand.

Each fraction is made up of two numbers:

$$\frac{\text{NUMERATOR}}{\text{DENOMINATOR}}$$

When adding fractions, arrange them either in vertical form as whole numbers or in linear form. In either case, the same procedure must be followed. Fractions cannot be added unless they have the same denominator!

If the fractions have the same denominators, add the numerators and put the sum over the denominator.

Example 1: Add $\frac{1}{9}$ and $\frac{7}{9}$.

$$\frac{1}{9} + \frac{7}{9} = \frac{1+7}{9} = \frac{8}{9}$$

Example 2: Find the sum of $\frac{6}{11}$ and $\frac{3}{11}$.

$$\frac{6}{11}$$
$$+\frac{3}{11}$$
$$\frac{6+3}{11} = \frac{9}{11}$$

If the fractions have different denominators, equivalent fractions with a common denominator must be created. To find the common denominator, take the largest denominator and then make multiples of that number (1 × the number, then 2 × the number, and so on) until you have a number that each of the denominators can be divided into evenly.

Example 3:   Add $\frac{1}{4}$, $\frac{2}{5}$, and $\frac{1}{8}$.

$$\frac{1}{4} + \frac{2}{5} + \frac{1}{8}$$

Here, 8 is the largest denominator. Four will divide evenly into 8; however, 5 will not. So take 2 × 8 or 16. Five will not divide evenly into 16, so try 3 × 8 = 24, then 4 × 8 = 32, then 5 × 8 = 40. This is the first multiple of 8 that 5 will divide into evenly. The common denominator, then, is 40.

Equivalent fractions must then be made with 40 as their denominators. An equivalent fraction is one that has the same value as another fraction but has a different denominator; therefore, it will also have a different numerator. A fraction can be made equivalent by multiplying both numerator and denominator by the same number. In this example, the number you use to multiply is the one needed to make the denominator 40.

$$\frac{1 \times 10}{4 \times 10} + \frac{2 \times 8}{5 \times 8} + \frac{1 \times 5}{8 \times 5} = \frac{10}{40} + \frac{16}{40} + \frac{5}{40}$$

Once the fractions have the common denominator, add the numerators.

$$\frac{10 + 16 + 5}{40} = \frac{31}{40}$$

The last step of each problem is to see if the fraction can be reduced. Many times (as in our first example) it cannot be reduced. There are two ways that fractions can be reduced.

Example 4:   Add $\frac{7}{18}$ and $\frac{5}{18}$.

$$\frac{7}{18} + \frac{5}{18} = \frac{7 + 5}{18} = \frac{12}{18}$$

In this example, the same factor, 6, can be divided evenly into the numerator and denominator.

$$\frac{12}{18} = \frac{12 \div 6}{18 \div 6} = \frac{2}{3}$$

Example 5: Find the sum of $\frac{3}{5}$, $\frac{19}{20}$, and $\frac{9}{10}$.

$$\frac{3}{5} = \frac{3 \times 4}{5 \times 4} = \frac{12}{20}$$

$$\frac{19}{20} = \frac{19}{20}$$

$$+ \frac{9}{10} = \frac{9 \times 2}{10 \times 2} = \frac{18}{20}$$

$$\frac{12 + 19 + 18}{20} = \frac{49}{20}$$

This answer has a fraction with the numerator larger than the denominator. Taking multiples of the denominator out of the numerator allows them to be written as whole numbers. The remainder is left as a fraction.

$$\frac{49}{20} = \frac{40}{20} + \frac{9}{20} = 2\frac{9}{20}$$

So a fraction can be reduced by taking common factors out of the denominator and numerator or by removing multiples of the denominator from the numerator. In some cases, both procedures are done.

With mixed numbers (a whole number and a fraction), work with the fractions and then with the whole numbers.

Example 6: Add $3\frac{3}{4}$ and $5\frac{5}{6}$.

$$\frac{3}{4} + \frac{5}{6} = \frac{3 \times 3}{4 \times 3} + \frac{5 \times 2}{6 \times 2}$$

$$= \frac{9}{12} + \frac{10}{12}$$

$$= \frac{9 + 10}{12} = \frac{19}{12}$$

$$= 1\frac{7}{12}$$

$$3 + 5 + 1\frac{7}{12} = 9\frac{7}{12}$$

A second way of working with mixed numbers is to convert the numbers to fractions, add, and then convert the answer back to a mixed number. The example given can be solved in this manner:

$$3\frac{3}{4} + 5\frac{5}{6}$$

$$= \left(\frac{12}{4} + \frac{3}{4}\right) + \left(\frac{30}{6} + \frac{5}{6}\right)$$

$$= \frac{15}{4} + \frac{35}{6}$$

$$= \frac{45}{12} + \frac{70}{12}$$

$$= \frac{45 + 70}{12} = \frac{115}{12} = \frac{108}{12} + \frac{7}{12} = 9\frac{7}{12}$$

Either way is correct.

Keep in mind the following guidelines when adding common fractions:

- Find the lowest common denominator.

- Make equivalent fractions with lowest common denominators.

- Reduce the answer to lowest terms.

PRACTICAL PROBLEMS

Note: In problems 1–12, add the fractions and mixed numbers.

1. $\dfrac{5}{13} + \dfrac{3}{13}$ _____

2. $\dfrac{5}{9} + \dfrac{1}{9} + \dfrac{2}{9}$ _____

3. $\dfrac{1}{4} + \dfrac{1}{8}$ _____

4. $\dfrac{2}{9} + \dfrac{2}{5}$ _____

5. $\dfrac{3}{5}$ 6. $\dfrac{1}{6}$ 7. $\dfrac{4}{5}$ 8. $\dfrac{5}{18}$ 9. $22\dfrac{5}{16}$

$\dfrac{1}{5}$ $\dfrac{4}{9}$ $\dfrac{3}{10}$ $\dfrac{3}{4}$ $7\dfrac{3}{4}$

$\dfrac{4}{5}$ $\dfrac{2}{3}$ $\dfrac{6}{25}$ $\dfrac{1}{3}$ $\dfrac{1}{8}$

$+\dfrac{3}{5}$ $+\dfrac{1}{2}$ $+\dfrac{7}{50}$ $\dfrac{5}{6}$ $+15\dfrac{1}{2}$

$+\dfrac{4}{9}$

10.   ¾ inch + ⅛ inch + ⁵⁄₈₄ inch + ⅜ inch + ⁷⁄₃₂ inch   _____

11.   ⁹⁄₁₆ inch + ⅝ inch + ½ inch + ¹⁄₃₂ inch + ¼ inch   _____

12.   2⅓ pounds + 15⅘ pounds + 7 pounds + 6¾ pounds   _____

13.   Find in inches the width of the galvanized sheet steel needed to make this
      duct.   _____

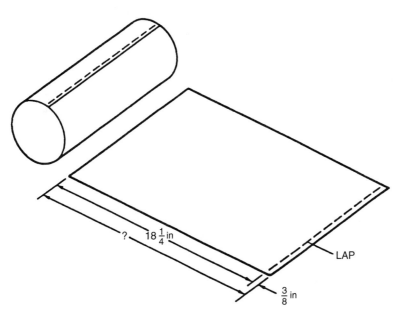

14. What is the center-to-center distance of the two bolt holes on this conduit pipe strap?   _____

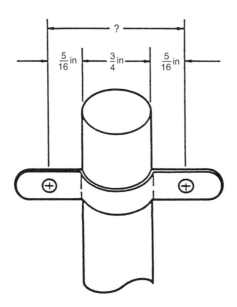

15. This filter-drier is to be added to a refrigerant circuit. What is the length this adds to the tubing?   _____

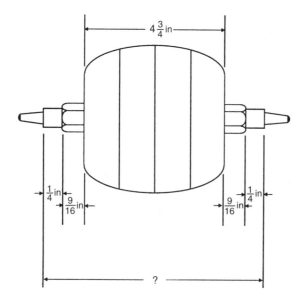

16. A fuel line must pass through this wall. Find in inches the length of the hole
     through the wall.   _____

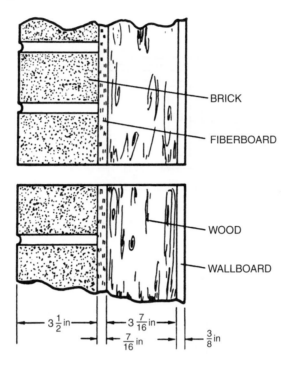

BRICK

FIBERBOARD

WOOD

WALLBOARD

$3\frac{1}{2}$ in $3\frac{7}{16}$ in

$\frac{7}{16}$ in $\frac{3}{8}$ in

17. A defective section of soft copper tubing must be replaced. A $21\frac{5}{16}$-inch
     section is cut out. The tube to replace the defective section will overlap
     $\frac{3}{8}$ inch at each end. How many inches long is the piece of new tubing?   _____

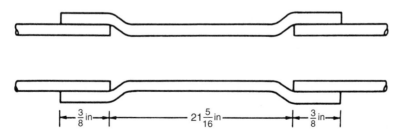

$\frac{3}{8}$ in $21\frac{5}{16}$ in $\frac{3}{8}$ in

18. What is the length of window seal gasket needed for this air-conditioning unit?   _____

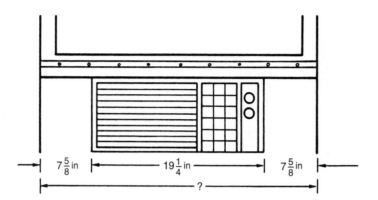

$7\frac{5}{8}$ in $19\frac{1}{4}$ in $7\frac{5}{8}$ in

?

19. A worn magnetic gasket around a refrigerator door must be replaced. Find the length of gasket needed.   _____

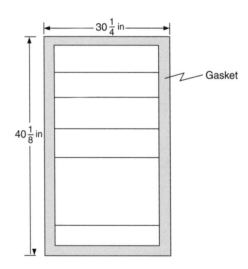

$30\frac{1}{4}$ in

$40\frac{1}{8}$ in

Gasket

20. A cooling fan is connected to the driving motor by a V-belt. How many inches long is the V-belt? _____

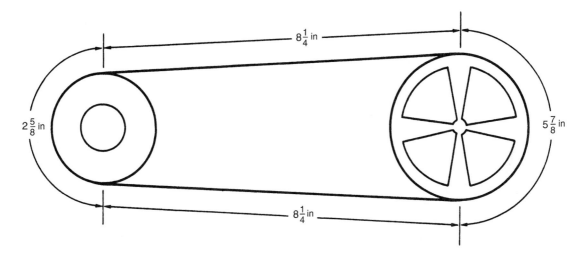

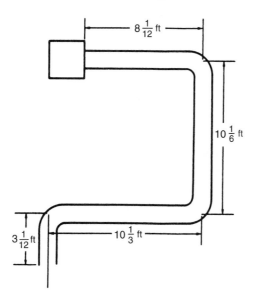

21. Find in feet the distance the air flows through this duct. _____

22. A conduit for the power wires of an air-conditioning unit is put together. How long is the conduit? _____

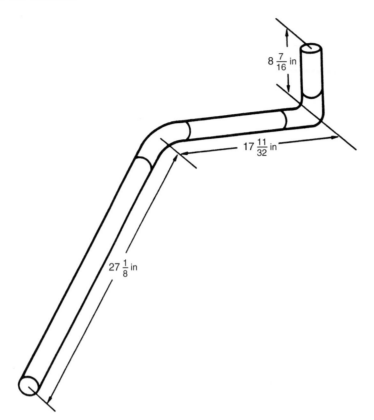

$8\frac{7}{16}$ in

$17\frac{11}{32}$ in

$27\frac{1}{8}$ in

23. A technician had to work on a single job on three different days. Monday, 6¾ hours were spent on the job; Tuesday, 4½ hours were spent on the job; and Thursday, 6 hours were spent on the job. What was the total time spent on the job? _____

24. A technician needs to drill a hole through a wall for a ½-inch fuel line. There should be a ¹⁄₃₂-inch clearance on each side of the tube. What should be the diameter of the drill? _____

25. A piece of ducting must be custom-made. It must fill a gap 25⁷⁄16 inches long.
     One end of the duct must also fit into the next piece of ducting 2⅜ inches.
     What is the total length this piece of ducting must be?   _____

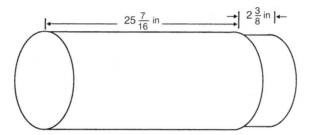

Unit 7 SUBTRACTION OF COMMON FRACTIONS

BASIC PRINCIPLES OF SUBTRACTION OF COMMON FRACTIONS

- Study subtraction of denominate numbers in Section I of the Appendix.

Subtracting fractions is almost like addition. You find the common denominator and make equivalent fractions. Subtract the numerators.

Example 1: Subtract $\frac{5}{18}$ from $\frac{8}{9}$.

$$\frac{8}{9} - \frac{5}{18} = \frac{8 \times 2}{9 \times 2} - \frac{5}{18}$$

$$= \frac{16 - 5}{18} = \frac{11}{18}$$

When subtracting mixed numbers, there may be times when a larger fraction is being subtracted from a smaller fraction.

Example 2: Find $5\frac{1}{3} - 3\frac{3}{5}$.

$$5\frac{1}{3} - 3\frac{3}{5} = 5\frac{1 \times 5}{3 \times 5} - 3\frac{3 \times 3}{5 \times 3}$$

$$= 5\frac{5}{15} - 3\frac{9}{15}$$

You cannot subtract $\frac{9}{15}$ from $\frac{5}{15}$. So what you do in this case is take 1 from the whole number 5. This leaves 4. Convert the 1 to an equivalent fraction with a denominator of 15. The numerator is also 15. Now add this fraction to the $\frac{5}{15}$. Then $\frac{9}{15}$ can be subtracted from this new fraction. This process is sometimes called borrowing.

$$5\frac{5}{15} - 3\frac{9}{15}$$

$$= 4\frac{15 + 5}{15} - 3\frac{9}{15}$$

$$= 4\frac{20}{15} - 3\frac{9}{15}$$

Now subtract the whole numbers and also subtract the fractions.

$$4\frac{20}{15} - 3\frac{9}{15}$$

$$= 1\frac{20 - 9}{15} = 1\frac{11}{15}$$

Use the following guidelines when subtracting common fractions:

- Find the lowest common denominator.

- Make equivalent fractions with lowest common denominators.

- You may have to borrow to complete the subtraction.

- Reduce the answer to lowest terms.

An equally valid way to subtract two mixed numbers is to convert the numbers totally to fractions, subtract, and then convert the answer to a mixed number.

Example 3: Find $5\frac{1}{3} - 3\frac{3}{5}$.

$$5\frac{1}{3} - 3\frac{3}{5}$$

$$= \frac{15 + 1}{3} - \frac{15 + 3}{5}$$

$$= \frac{16}{3} - \frac{18}{5}$$

$$= \frac{16 \times 5}{3 \times 5} - \frac{18 \times 3}{5 \times 3}$$

$$= \frac{80}{15} - \frac{54}{15}$$

$$= \frac{26}{15} = 1\frac{11}{15}$$

PRACTICAL PROBLEMS

Note: In problems 1–14, subtract the fractions and mixed numbers.

1. $\dfrac{5}{7} + \dfrac{2}{7}$ _____

4. $6\dfrac{5}{9} - 2\dfrac{1}{3}$ _____

2. $\dfrac{3}{4} + \dfrac{5}{16}$ _____

5. $7\dfrac{1}{5} - 2\dfrac{3}{7}$ _____

3. $5\dfrac{4}{5} - 2\dfrac{1}{5}$ _____

6. $\dfrac{7}{16}$ $-\dfrac{5}{16}$

7. $\dfrac{2}{3}$ $-\dfrac{4}{25}$

8. $3\dfrac{4}{5}$ $-1\dfrac{1}{3}$

9. $2\dfrac{2}{5}$ $-1\dfrac{7}{8}$

10. $4\dfrac{1}{16}$ $-\dfrac{4}{5}$

11. ¼ inch − ¹⁄₁₆ inch   _____

12. ⁹⁄₃₂ inch − ⅛ inch   _____

13. 9⅞ inches − 7⅝ inches   _____

14. 15¼ inches − 9⅞ inches   _____

15. A piece of polyvinylchloride (PVC) tubing, 3⅜ feet long, is needed for a drain of an air-conditioning unit. The piece is cut from an 8-foot-long coil of tubing. How much tubing is left?   _____

16. How many inches does this window air conditioner extend inside the window?   _____

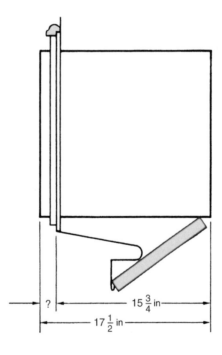

17. This air-conditioning condenser unit is mounted on the top of a flat-roofed building. Find in inches clearance A. _____

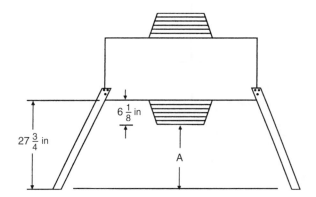

18. In one day, a technician works 9½ hours and finishes two jobs. It takes 3¾ hours to finish the first job. How long does it take to finish the second job? _____

19. Find in inches the drop in height for this fuel line. _____

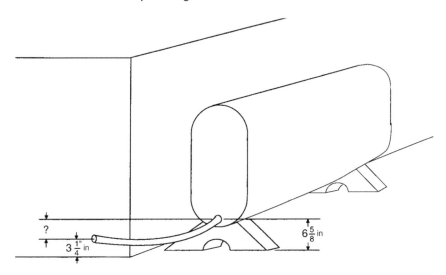

20.   A spray nozzle humidifier is installed in this air duct.

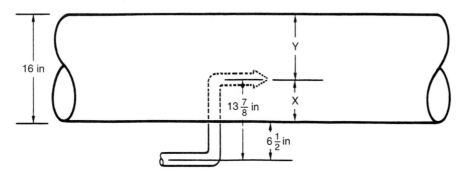

a. Find in inches dimension X.   _____

b. Find in inches dimension Y.   _____

21.   Determine dimension A on this refrigerator.   _____

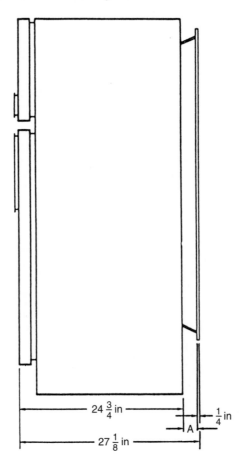

22. This vent cover is fitted into a duct.

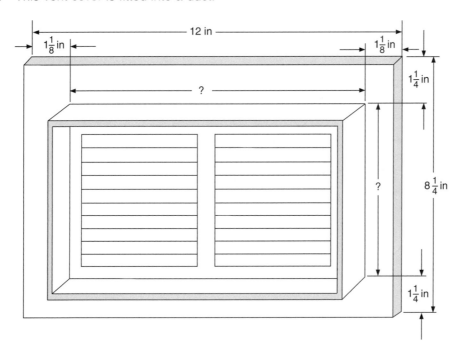

    a. How many inches high must the duct be?   _____

    b. How many inches wide must the duct be?   _____

23. What is the length of the finned section of pipe in this baseboard hot water heating system?   _____

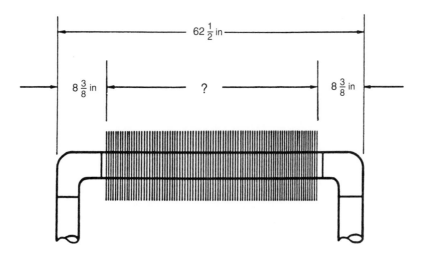

24. Five air-conditioning units are checked and recharged with refrigerant R-134a. A full cylinder of R-134a contains 25 pounds of refrigerant. These amounts are taken from a full cylinder: 3½ pounds, 2⅓ pounds, 4⅔ pounds, 2⅘ pounds, and 5¹⁄16 pounds. How much R-134a is left?

_____

25. How far away from the wall is the end of the heater pictured below?

_____

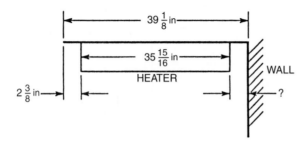

Unit 8 MULTIPLICATION OF COMMON FRACTIONS

BASIC PRINCIPLES OF MULTIPLICATION OF COMMON FRACTIONS

- Study multiplication of denominate numbers in Section I of the Appendix.

Multiplying fractions is easier than adding or subtracting them. It is best to put the fractions in a linear form. Then it is simply a case of multiplying the numerators and also multiplying the denominators.

Example 1: Multiply $\frac{2}{3}$ by $\frac{4}{5}$.

$$\frac{2}{3} \times \frac{4}{5} = \frac{2 \times 4}{3 \times 5} = \frac{8}{15}$$

Once again, the last thing to do is to check if the answer can be reduced. The reducing can take place in an interim step: before multiplying, see if there is any number that can divide into one numerator and one denominator. This is called cancelling. In the example below, the factor is 3.

Example 2: Find the product of $\frac{2}{3}$ and $\frac{9}{11}$.

$$\frac{2}{\overset{}{\underset{1}{3}}} \times \frac{\overset{3}{9}}{11} = \frac{2 \times 3}{1 \times 11} = \frac{6}{11}$$

When multiplying, do not multiply mixed numbers. Change all mixed numbers to fractions before multiplying. You will probably have to change the answer back to a mixed number when you are done.

Example 3: Find $1\frac{1}{3}$ times $2\frac{5}{6}$.

$$1\frac{1}{3} \times 2\frac{5}{6} = \frac{\overset{2}{4}}{3} \times \frac{17}{\underset{3}{6}}$$

$$= \frac{2 \times 17}{3 \times 3} = \frac{34}{9}$$

$$= 3\frac{7}{9}$$

Follow these guidelines when multiplying common fractions:

- Remember, when multiplying fractions, you do not need to find the lowest common denominator.

- When cancelling, one of the numerators and one of the denominators must be divided by the same number.

- You do not have to cancel; cancelling just makes the problem easier because you work with smaller numbers.

- You may cancel more than once.

- Reduce the answer to lowest terms.

PRACTICAL PROBLEMS

Note: In problems 1–11, multiply the fractions and mixed numbers.

1. $\dfrac{1}{4}$

   $\times \dfrac{3}{4}$

2. $\dfrac{1}{3}$

   $\times \dfrac{3}{7}$

3. $\dfrac{2}{3}$

   $\times \dfrac{4}{5}$

4. $\dfrac{2}{7}$

   $\times \dfrac{49}{50}$

5. $4\dfrac{1}{5}$

   $\times \dfrac{2}{3}$

6. $2\dfrac{1}{4}$

   $\times 1\dfrac{2}{5}$

7. $1\dfrac{1}{35}$

   $\times 1\dfrac{3}{4}$

8. $\frac{1}{2}$ inch $\times \frac{1}{4}$ _____

9. $1\frac{1}{6}$ feet $\times \frac{1}{3}$ _____

10. $3\frac{1}{3}$ feet $\times 2\frac{1}{20}$ _____

11. $\frac{3}{5} \times 1\frac{1}{4}$ inches $\times 3\frac{1}{6}$ _____

12. Convert $2\frac{1}{4}$ feet to inches. _____

13. Convert $5\frac{1}{10}$ feet to inches. _____

14. The minimum safe bending radius of a certain polyethylene tube is the outside diameter times $3\frac{1}{2}$. The tube has an outside diameter of $\frac{3}{8}$ inch. What is the minimum bending radius? _____

15. A house is being built. The contractor states that the cost to install electric baseboard heat will be $2,480. The contractor also states that a forced air, heat pump system will be 2¼ times as much. How much will the heat pump system cost?

_____

16. What is the total distance between the centers of the first and last gas burners?

_____

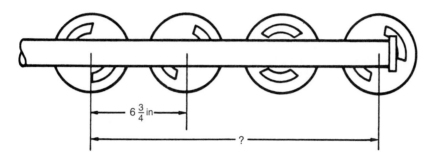

17. A supply plenum is fastened to the furnace with equally spaced screws. Find in inches dimension X.

_____

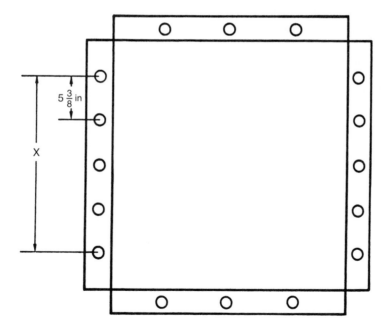

18. A fuel tank holds 281½ gallons. One gallon of *#2* fuel weighs 7¹⁄₁₀ pounds. How many pounds of fuel are in a full tank?   _____

19. Support straps are used for this duct. Each support strap is made from two pieces of metal each 21³⁄₈ inches long. How many inches of metal are needed for the straps?   _____

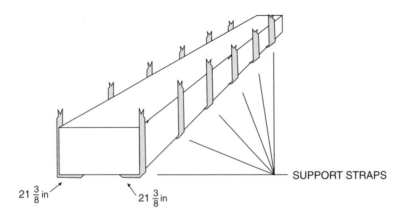

$21 \frac{3}{8}$ in $21 \frac{3}{8}$ in   SUPPORT STRAPS

20. Each hour an oil burner runs, the nozzle sprays ¾ gallon of fuel. If the burner runs for 2½ hours, how much fuel is sprayed into the burner?   _____

21. Washers that fit a ³⁄₁₆-inch bolt cost 1²⁄₉¢ each. What is the cost of 36 (3 dozen) washers?   _____

22. This condenser coil is made from tubing that has a ¼-inch outside diameter.
     How many inches of tubing are needed for the coil?   _____

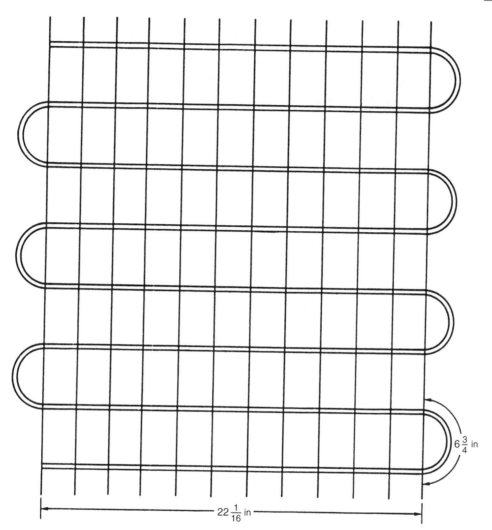

$6\frac{3}{4}$ in

$22\frac{1}{16}$ in

23. A technician spends 9 hours at work. Repairing a refrigerator takes ½ of the
     time and repairing a window air conditioner takes ⅓ of the time. The
     technician spends ¹⁄₁₈ of the time ordering parts and ⅑ of the time eating
     lunch. Find the number of hours the technician spends in each activity.

     a. repairing refrigerator   _____

     b. repairing air conditioner   _____

     c. ordering parts   _____

     d. eating lunch   _____

24. A technician is recovering refrigerant R-22 to be properly disposed. Each unit emptied produced $25\frac{5}{6}$ ounces of R-22. If 8 units are emptied, how many ounces of R-22 did the technician recover?

        _____

25. At 63°F, refrigerant R-22 has a weight of $2\frac{1}{5}$ pounds for each cu ft of the R-22 in gaseous form. A technician determines that the tubing from the evaporator back to the condenser has a volume of $\frac{1}{64}$ cu ft. If the technician recovers the R-22 in this tubing, how many pounds will be recovered?

        _____

Unit 9 DIVISION OF COMMON FRACTIONS

BASIC PRINCIPLES OF DIVISION OF COMMON FRACTIONS

• Study division of denominate numbers in Section I of the Appendix.

Division of fractions is very similar to multiplication of fractions with one exception. Take the divisor (the fraction after the ÷ sign) and invert it (make the numerator the denominator and the denominator the numerator). Then change the ÷ sign to a × sign and multiply the two fractions.

The tough part of a division of fractions problem is when you have a word problem. There is no ÷ sign, making it hard to tell which fraction to invert. A division problem can always be written as

$$\text{DIVIDEND} \div \text{DIVISOR} = \text{QUOTIENT}$$

You need to be able to decide which is the dividend and which is the divisor in a word problem. Two hints may help you. Many problems read something *divided by* something. You can just replace the words *divided by* with ÷. Other problems can be thought of as a number of equal things. It is the number of equal things or the value of the equal things (size, weight, length, and so on) that becomes the divisor. What these things are coming from is the dividend. Following these two hints will allow you to solve most division problems.

Example 1: Divide ⅜ by ⁵⁄₇.

$$\frac{3}{8} \div \frac{5}{7} = \frac{3}{8} \times \frac{7}{5} = \frac{3 \times 7}{8 \times 5}$$
$$= \frac{21}{40}$$

When dividing with mixed numbers, always convert the mixed numbers to fractions before dividing.

Example 2: Divide 1⅟₇ by 2⅟₉.

$$1\frac{1}{7} \div 2\frac{1}{9} = \frac{7+1}{7} \div \frac{18+1}{9} = \frac{8}{7} \div \frac{19}{9}$$
$$= \frac{8}{7} \times \frac{9}{19} = \frac{72}{133}$$

Be careful not to cancel until after you have inverted!

Example 3: Find $1\frac{1}{3}$ divided by $1\frac{1}{2}$.

$$1\frac{1}{3} \div 1\frac{1}{2} = \frac{3+1}{3} \div \frac{2+1}{2} = \frac{4}{3} \div \frac{3}{2}$$

$$= \frac{4}{3} \times \frac{2}{3}$$

$$= \frac{8}{9}$$

Notice you *cannot* cancel.

Use the following guidelines when dividing common fractions:

- Invert the fraction after the $\div$ sign and change the $\div$ sign to a $\times$ sign.

- After inverting, treat the problem as a multiplication problem.

- Do not cancel before inverting.

PRACTICAL PROBLEMS

Note: In problems 1–11, divide the fractions and mixed numbers.

1. $\frac{2}{5} \div \frac{7}{9}$ _____

2. $3\frac{1}{5} \div \frac{4}{7}$ _____

3. $3 \div 1\frac{2}{3}$ _____

4. $1\frac{7}{25} \div \frac{2}{5}$ _____

5. $5\frac{4}{9} \div 8\frac{2}{3}$ _____

6. $6\frac{2}{9} \div 1\frac{17}{18}$ _____

7. $\frac{1}{8}$ inch $\div$ 2 _____

8. $\frac{3}{4}$ hour $\div$ 3 _____

9. $\frac{3}{7}$ week $\div \frac{1}{4}$ _____

10. $6\frac{2}{3}$ yards $\div \frac{2}{3}$ yard _____

11. $142\frac{1}{3}$ Btu $\div 5\frac{1}{4}$ hours _____

12. Air duct straps are made from a 1-inch-wide strip of metal. The metal is $246\frac{3}{4}$ inches long. Each strap is $17\frac{5}{8}$ inches long. How many straps can be made? _____

13. Air-conditioning systems are charged with refrigerant R-134a. The cylinder of refrigerant used contains $24\frac{1}{2}$ pounds. Each system uses $1\frac{3}{4}$ pounds of R-134a. How many systems can be charged from the $24\frac{1}{2}$-pound cylinder? _____

14. A section of a condenser measures 4¼ inches and has 51 fins. Some of the fins are bent and must be straightened with a fin comb. The fin comb must have the same number of fins per inch as the condenser. How many fins per inch does the fin comb have? _____

15. Strips of metal are cut from a 3-foot by 8-foot (96-inch) sheet. Each strip is 3 feet long and 19¾ inches wide. Find the number of complete strips that can be made from this sheet. _____

16. These 5 air-conditioning condensing units are mounted next to each other on a roof. Each unit is equal in size. What is the width of each unit in inches? _____

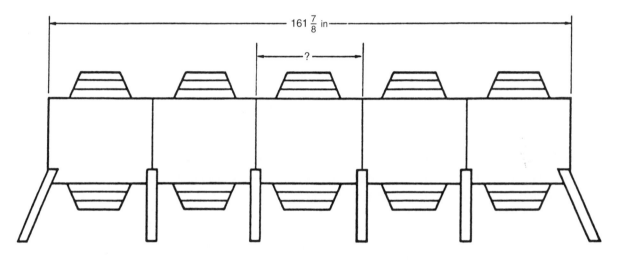

17. Washers are placed on a leveling screw to give it support. The washers are *18* gauge and are ³⁄₆₄ inch thick. How many washers are needed to fill the space? _____

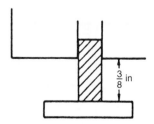

18. A piece of *12*-gauge sheet metal is $^{7}/_{64}$ inch thick. A stack of sheet metal is 17$^{1}/_{2}$ inches thick. How many sheets are in the stack?

_____

19. A certain house has floor joists every 1$^{1}/_{3}$ feet. The heating duct for the house runs under the floor hanging from straps that are attached to some of the joists. The distance between the support straps is 4 feet. How many joists have straps attached to them?

_____

20. Duct tape is used to seal joints in duct insulation. A roll of duct tape contains 360 inches of tape. A 10-inch round duct with a 1-inch layer of insulation on it will need a piece of tape 37$^{3}/_{4}$ inches long to cover the seam and overlap itself. How many seams will one roll of duct tape cover?

_____

21. An installer can work 6$^{3}/_{4}$ hours each day at a worksite. The installer will need 30$^{3}/_{8}$ hours to do a complete installation of a heating and cooling system. How many days will the installer be at the job site?

_____

22. Four window air-conditioning units are shrink-wrapped together. The package weighs 249$^{3}/_{4}$ pounds. How much does each window unit weigh?

_____

23. A technician recovered 22$^{1}/_{2}$ pounds of refrigerant R-12 for disposal by emptying $^{5}/_{6}$ of a pound of R-12 from each of a number of portable dehumidifier units. How many dehumidifier units did the technician empty?

_____

24. A humidifier added 2$^{5}/_{8}$ quarts of water to a forced air system each 10$^{1}/_{2}$ hours of operation. How much water was added to the air each hour?

_____

25. Single wall smoke pipe is used to connect the oil burner to the chimney flue. The distance from the burner to the vent is 157$^{1}/_{2}$ inches. If 1$^{1}/_{2}$ inches are allowed on each pipe for fitting, how many lengths of pipe are needed?

_____

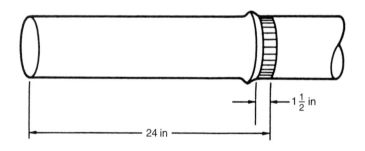

Unit 10 COMBINED OPERATIONS WITH COMMON FRACTIONS

BASIC PRINCIPLES OF COMBINED OPERATIONS WITH COMMON FRACTIONS

- Review and apply the principles of addition, subtraction, multiplication, and division of common fractions to these problems.

- Read each problem. Decide which operation must be performed to solve the problem. Perform that operation.

- Some problems may require the application of more than one type of operation to solve the problem. As a result, it may take more than one step to find the answer.

PRACTICAL PROBLEMS

1. $2\frac{1}{5}$
    $+3\frac{5}{9}$

2. $5\frac{2}{3}$
    $7\frac{7}{8}$
    $+2\frac{5}{6}$

3. $4\frac{5}{6}$
    $-2\frac{1}{2}$

4. $4\frac{1}{5}$
    $-3\frac{6}{7}$

5. $\frac{5}{6}$
    $\times\frac{1}{4}$

6. $5\frac{5}{6}$
    $\times2\frac{2}{7}$

7. $\frac{2}{3} \div \frac{5}{7}$ _____

8. $2\frac{4}{9} \div 2\frac{1}{5}$ _____

9. $\frac{1}{6}$ yard $+ \frac{3}{8}$ yard _____

10. $\frac{7}{8}$ inch $- \frac{5}{32}$ inch _____

11. $\frac{15}{16}$ pound $\times \frac{4}{5}$ _____

12. $4\frac{1}{6}$ yards $\div 3\frac{1}{3}$ _____

13.   An oil burner is installed in a basement. The floor of the basement slopes and
      the oil burner must be leveled. What is the height from the floor to the top of
      the burner after leveling?   _____

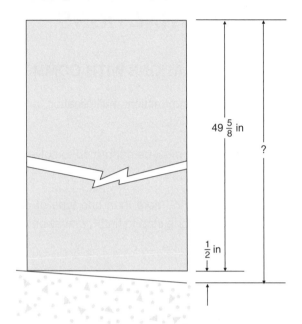

14.   A technician uses *#10* wire for a repair job. Before starting the job, the roll has
      78½ feet of wire on it. The technician uses ⅓ of the roll. How many feet of
      wire does the technician use?   _____

15.   The nozzle and electrodes of an oil burner extend from the end of the burner
      gun. Find in inches dimension X.   _____

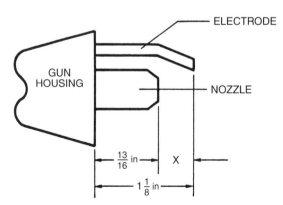

16. What is the total length of this furnace and air-conditioning system?   _____

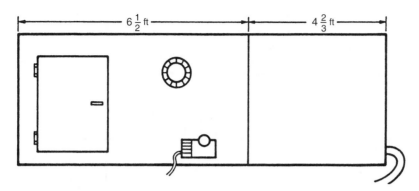

17. The tubes going to and from an air-conditioning condensing unit must pass through a wall. The tubes have diameters of $1\frac{7}{8}$ inches and $\frac{3}{4}$ inch. What is the smallest diameter hole that can be used?   _____

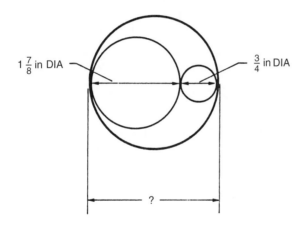

18. At the start of a workday, a cylinder of refrigerant R-134a has $22\frac{1}{2}$ pounds in it. A technician works on 4 air-conditioning units and uses all of the refrigerant. On the first unit, $\frac{1}{4}$ of the refrigerant is used, and $\frac{1}{12}$ of the refrigerant is used on the second unit. Then $\frac{1}{6}$ of the refrigerant is used on the third unit and $\frac{1}{2}$ on the fourth. Find the number of pounds of refrigerant R-134a used on each unit.

   a. Unit 1   _____

   b. Unit 2   _____

   c. Unit 3   _____

   d. Unit 4   _____

19.   Find in inches the diameter of this duct.   _____

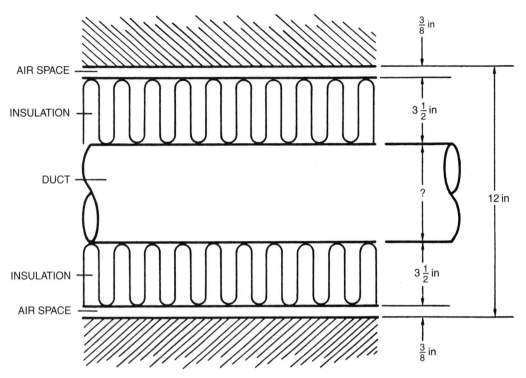

20.   A circular duct has an outside diameter of 7½ inches. Insulation that is 1⅛ inches thick is wrapped around the duct. What is the diameter of the insulated duct?   _____

21. What is the height of the finished ceiling in this room? _____

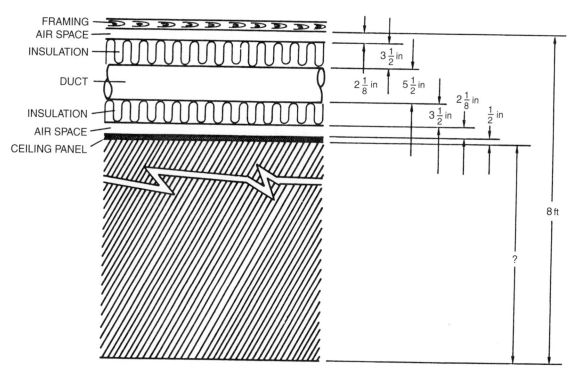

22. A piece of duct measures 24 inches. When 2 pieces are fitted together, 1½ inches are allowed on each duct for fitting. On a job, 27½ pieces of ducting are used. Find in inches the length of the finished duct. _____

23. Two filler boards of equal width are used to install this window air-conditioning casing. Find the width of each board. _____

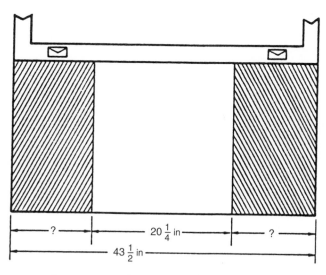

24. This ceiling register has equally spaced openings.

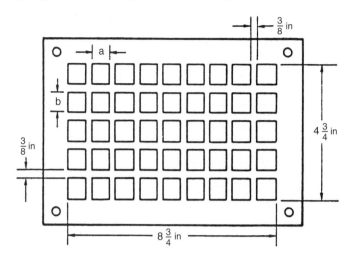

    a. Find in inches the width of each opening.

    b. Find in inches the height of each opening.

25. A ventilating system is to be installed in a building. It will require 400 hours of work to install it. Each worker takes a total of ½ hour to get tools out in the morning and put them away at night. Each worker also takes a total of ¼ hour to get to the workplace and back again each day. Workers are also given a ¼-hour break in the morning and another in the afternoon.

    a. For an 8-hour workday, how much time is actually spent installing the ventilating system?

    b. If 5 people are in the work crew, how many hours of work are done each day?

    c. How many days will be needed to complete the job with a 5-person crew?

Decimal Fractions

Unit 11 ADDITION OF DECIMAL FRACTIONS

BASIC PRINCIPLES OF ADDITION OF DECIMAL FRACTIONS

•   Study addition of denominate numbers in Section I of the Appendix.

A decimal number is a number with a decimal point in it. The decimal point separates the whole number part of the number from the fraction part. The whole number part is to the left of the decimal point, while the fraction part is to the right. Each number position of the fraction part has a name, just as each position of a whole number has a name. So the number 123.4567809 has 123 as the whole number part and 4567809 as the fraction part. Four is in the tenths position, 5 in the hundredths position, 6 in the thousandths position, 7 in the ten-thousandths position, 8 in the hundred-thousandths position, 0 in the millionths position, and 9 in the ten-millionths position.

Adding decimal numbers is just like adding whole numbers, with one additional concern. When writing the numbers, line up the decimal points.

Example 1:   Find the sum of 233.45, 18.9, and 506.807.

$$\begin{array}{r} 233.45 \\ 18.9 \\ +506.807 \\ \hline \end{array}$$

Notice that no zeros are written at the end of the fraction parts of the numbers. An exception to this is when dealing with money. Add the columns just as was done with whole numbers. Carry numbers just as was done with whole numbers. Place the decimal point directly under where it is in the column.

$$\begin{array}{r} {\scriptstyle 1\,2} \\ 233.45 \\ 18.9 \\ +506.807 \\ \hline 759.157 \end{array}$$

When adding a whole number with a decimal, the question that arises is where is the decimal point in the whole number. For a whole number, the decimal point is to the right of the unit number. This allows you to properly line up the decimal points.

Remember: When adding decimal fractions, always make sure that the decimal points are lined up one under the other.

Calculators can be used to solve problems with decimal fractions much more easily than with regular fractions. Once again, it is important to know if the answer that you get on your calculator is close to what you estimate the answer to be. Round the numbers off and solve the problem to give you a number that should be close to the actual answer. Let us look at two examples.

Example 2:   Estimating the problem given above:

$$
\begin{array}{r}
233.45 \\
18.9 \\
+506.807 \\
\end{array}
\qquad \Rightarrow \qquad
\begin{array}{r}
200 \\
20 \\
+500 \\
\hline
720 \\
\end{array}
$$

So the estimated answer is 720. Using the calculator, entering 233.45 + 18.9 + 506.807 = gives 759.157, which is close to 720.

Example 3:   Three shims are being used to align a compressor and motor. The thicknesses of the shims are 0.43 inches, 0.28 inches, and 0.03 inches. What size single shim can replace these three?

We are looking for the total, so we will add these three numbers. Estimating the answer:

$$
\begin{array}{r}
0.43 \\
0.28 \\
+0.03 \\
\end{array}
\qquad \Rightarrow \qquad
\begin{array}{r}
0.4 \\
0.3 \\
+0.03 \\
\hline
0.73 \\
\end{array}
$$

Entering .43 + .28 + .03 = in a calculator gives .74, which is close to our estimated answer. Since all of the measurements were inches, the answer is 0.74 inches.

Even though this problem has small numbers, we round them off to one numeral each, as we did with large numbers.

PRACTICAL PROBLEMS

Note: In problems 1–12, add the decimal fractions.

1.   214.71
      172.55
    +187.37

4.   1,731.862 meters
          3.14 meters
    +   92.67 meters

2.   0.913
       8.047
     +76.465

5.   0.812 cu in
       960.245 cu in
        37.043 cu in
     +251.3   cu in

3.   54.
      330.713
    +   8.92

6.   121.44 sq m
        40.08 sq m
         5.91 sq m
     +636.36 sq m

7.   58.81 + 49.97 + 36.54

8.   32.987 + 1,993.25 + 520.09   _____

9.   0.2623 + 1,750.97 + 6.844   _____

10.   29.97 liters + 0.3484 liter + 1.822 liters   _____

11.   225 pounds + 598.17 pounds + 1.982 pounds   _____

12.   409.65 gallons + 0.644 gallons + 79.7 gallons + 69.286 gallons   _____

13.   In a normal refrigeration cycle, the vapor pressure for refrigerant R-134a
     entering the compressor is 19.3 pounds per sq in gauge reading (psig). The
     condensing pressure is 150.5 psig higher than the vapor pressure. What is
     the condensing pressure for refrigerant R-134a?   _____

14.   Absolute pressure = Gauge pressure + Atmospheric pressure
     A pressure gauge in a refrigerator system is reading 119.6 pounds per sq in
     gauge pressure when the atmospheric pressure is 14.7 pounds per sq in (psi).
     What is the absolute pressure of the refrigerant in that system?   _____

15. The gasket on this refrigerator is worn and no longer fills the gap. The gasket is to be replaced. Find the thickness of the new gasket.   _____

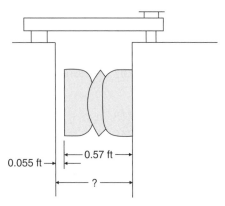

0.055 ft →   —0.57 ft —→

?

16. A section of a hot water heating system must be suspended. The section is made from one-inch copper tubing weighing 8.515 pounds. The water inside the tubing weighs 4.6558 pounds. What is the total weight that must be supported in this section?   _____

17. In a refrigeration cycle, the refrigerant gains heat in the evaporator and in the suction line. In a certain refrigeration system, R-134a gains 67.8 British thermal units per pound (Btu/lb) in the evaporator. It then gains 2.5 Btu/lb in the suction line. Find in Btu/lb the total heat gained by the R-134a.   _____

18. In a parallel electrical circuit, the total current is the sum of the individual currents. An air conditioner has a blower and a compressor that are wired in a parallel circuit. The blower draws 1.042 amperes and the compressor draws 11.95 amperes of current. When both are operating, what is the total current drawn?   _____

19. What is the outside diameter of this tube?   _____

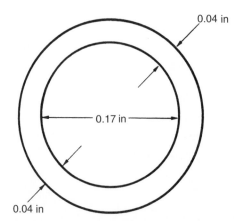

0.04 in

0.17 in

0.04 in

20. A technician uses these items in fixing a hot water heating system:

ITEM	COST
acid core solder	$2.49
solder paste	$0.59
stop valve	$2.85
90° elbow	$0.43
$\frac{1}{2}$-inch copper tubing	$5.34

Find the total cost of the items used.   _____

Note: Use this information and diagram for problems 21 and 22. The *R value* of a substance gives the resistance of a substance to heat flow. The larger the R value, the more resistant the substance is to heat flow and heat loss.

R VALUES FOR CERTAIN MATERIALS

MATERIAL	R VALUE
Air space ($3\frac{1}{2}$ inches)	0.97
Brick veneer	0.56
Fiberglass insulation ($3\frac{1}{2}$ inches)	11.00
Wallboard (interior)	0.79
Wood siding and building paper	0.86

The total resistance is the sum of each substance's resistance.

21. What is the total R value for this wall? _____

22. Fiberglass insulation, $3\frac{1}{2}$ inches thick, is placed in the air space in the wall. Find the R value. _____

23. A partially filled cylinder of refrigerant R-124 weighs 47.3 pounds. Another 15.5 pounds of R-124 are put into the cylinder. How much does the cylinder now weigh? _____

24. At 86 degrees Fahrenheit (°F) and 111.83 psi absolute, R-134a boils (it changes from a liquid to a gas). One pound of liquid R-134a occupies 0.0135 cu ft of volume. When it changes to a gas, it requires 0.4132 cu ft more space. What volume will 1 pound of R-134a occupy as a gas at 86°F and 111.83 psi? _____

25. Upon being started, a motor draws additional current until the motor is running. A large fan motor draws 3.45 amperes of current when running. When starting, the fan motor draws an additional 9.27 amperes of starting current. What is the total current drawn by the fan motor while starting? _____

Unit 12 SUBTRACTION OF DECIMAL FRACTIONS

BASIC PRINCIPLES OF SUBTRACTION OF DECIMAL FRACTIONS

- Study subtraction of denominate numbers in Section I of the Appendix.

To subtract decimals, once again line up the decimal points. Then subtract just as was done with whole numbers. You may have to borrow as was done with whole numbers.

Example: Subtract 8.46 from 25.3.

$$
\begin{array}{r}
25.3 \\
-\ 8.46 \\
\hline
\end{array}
$$

Treat the blank after the 3 as a zero. Borrow from the 3 and make the 0 a 10. In this problem, you will also have to borrow from the 5 to make a 12 above the 4 and from the 2 to make a 14 above the 8.

$$
\begin{array}{r}
14\ \ 12 \\
210 \\
25.3 \\
-\ 8.46 \\
\hline
16.84
\end{array}
$$

Just like in addition of decimals, the decimal point in the answer goes under the decimal points that were lined up.

To estimate this answer, we can either round the problem to

$$
\begin{array}{r}
25.3 \\
-\ 8.46 \\
\hline
\end{array}
\Rightarrow
\begin{array}{r}
30 \\
-\ 8 \\
\hline
22
\end{array}
\ \text{or}\
\begin{array}{r}
25 \\
-\ 8 \\
\hline
17
\end{array}
$$

In either case, we see that the answer will be about 20. Entering the problem into the calculator, $25.3 - 8.46 =$ gives an answer of 16.84, which is close to either estimate.

PRACTICAL PROBLEMS

Note: In problems 1–12, subtract the decimal fractions.

1.   249.83
     −126.41

2.   0.9366
     −0.1718

3.   3.754
     −2.46

4.   325.19 sq cm
     − 44.053 sq cm

5.   75.013 sq in
     −32.246 sq in

6.   582.476 liters
     −488.827 liters

7.   259.5 − 169.7   _____

8.   534.93 − 401.46   _____

9.   1,447.755 − 829.74   _____

10.   597.6 sq ft − 321.04 sq ft   _____

11.   4.5189 sq m − 0.2616 sq m   _____

12.   110 pounds − 67.219 pounds   _____

Note: Use this table for problems 13–15.

PROPERTIES OF CERTAIN REFRIGERANTS

REFRIGERANT	TEMPERATURE (in °F)	DENSITY (in lb/cu ft)	SATURATED VAPOR PRESSURE (in lb/sq in)
R-134a	5 °F 86 °F	0.517 2.344	23.772 111.828
R-124	5 °F 86 °F	0.374 1.732	12.994 64.585

13.   At 5°F, how much denser is R-134a than R-124?   _____

14.   What is the difference in the saturated vapor pressure for refrigerant R-134a at the two given temperatures?   _____

15.   How much denser is R-124 at 86°F than at 5°F?   _____

16. The bearing in an electric fan motor should be replaced if the rotor hits the stator. When assembled, the distance between the rotor and the stator is 0.03 inch. The bearing has had 0.00175 inch worn off. Find the distance remaining between the rotor and the stator.   _____

17. In order for a certain refrigerator door to close tightly, the refrigerator must be leveled up 0.24 inch using shims. If two shims are used, what is the thickness of the second shim?   _____

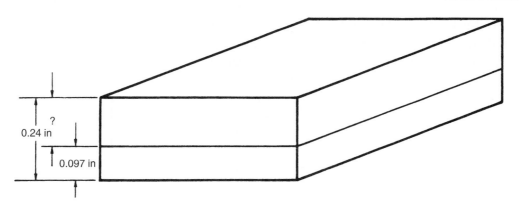

18. At 150°F dry air has a density of 0.0652 lb/cu ft. At 70°F it has a density of 0.075 lb/sq ft. What is the difference in density between air at 70°F and air at 150°F?   _____

19. Determine in psi the pressure drop between the ends of this duct.   _____

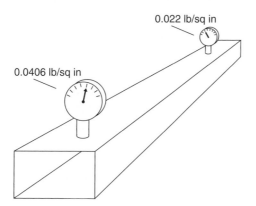

20.   At 5°F, 1 pound of refrigerant R-124 vapor has a volume of 2.6717 cu ft. When it is compressed into a liquid, it has a volume of 0.0108 cu ft. Find the change in volume between the vapor and liquid states.   _____

21.   Some refrigerants are a mixture of three different refrigerants, R-32, R-125, and R-134a. In 1 pound of such a mixture, R-32 and R-125 together weigh 0.706 pound. How much R-134a is in this mixture?   _____

22.   The R value of a substance gives the resistance of that substance to heat flow. The larger the R value, the more resistant the substance is to heat flow and heat loss. The total resistance of this wall is 17.85. The sum of the R values shown is 17.06. What is the missing R value?   _____

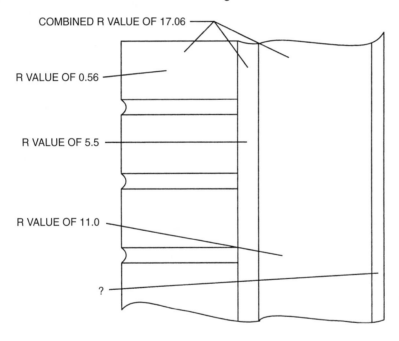

COMBINED R VALUE OF 17.06

R VALUE OF 0.56

R VALUE OF 5.5

R VALUE OF 11.0

?

23.   A partially filled cylinder of R-407c weighs 47.3 pounds. When the cylinder is empty, it weighs 12.6 pounds. What is the weight of the R-407c in the partially filled cylinder?   _____

24.   A bill is made out for a completed repair job. The bill is broken down into parts and labor. The total of the bill is $84.57. Parts cost $27.68. What is the cost for labor?   _____

25.   Absolute pressure = Gauge pressure + Atmospheric pressure
A pressure gauge is being checked for accuracy. The gauge is connected to a tank that has an absolute pressure of 742.11 psi. Atmospheric pressure is 14.7 psi. What should the gauge read?   _____

Unit 13 MULTIPLICATION OF DECIMAL FRACTIONS

BASIC PRINCIPLES OF MULTIPLICATION OF DECIMAL FRACTIONS

- Study multiplication of denominate numbers in Section I of the Appendix.

When multiplying decimals, do not worry about where the decimal points are. Line up the last numbers. Multiply the numbers just as was done with whole numbers.

Example: Multiply 51.3 by 0.469.

$$
\begin{array}{r}
51.3 \\
\times 0.469 \\
\hline
4617 \\
3078 \\
2052 \\
\hline
240597
\end{array}
$$

Now add the total number of places to the right of the decimal point in the numbers being multiplied together—one in the first number and 3 in the second number. Count that number of places from the right end of the answer and place the decimal point.

The answer 24.0597 has the decimal point in the proper location.

A whole number has 0 places to the right of the decimal point. When multiplying a decimal number times a whole number, the answer should have as many decimal places in it as the decimal number in the problem.

To solve this problem, estimate the answer first. If using a calculator, estimating the answer for this problem gives

$$
\begin{array}{r}
51.3 \\
\times 0.469 \\
\hline
\end{array}
\quad \Rightarrow \quad
\begin{array}{r}
50 \\
\times\ .5 \\
\hline
25.0
\end{array}
$$

Entering the problem into the calculator 51.3 × .469 = gives an answer of 24.0597, which is close to the estimate.

Remember: When multiplying decimal fractions, the decimals do not have to be lined up. The number of places to the right of the decimal point in the product is equal to the sum of the number of places to the right of the decimal point in both numbers being multiplied.

PRACTICAL PROBLEMS

Note: In problems 1–12, multiply the decimal fractions.

1. 503.6
    $\times$ 4.47

2.   3.594
    $\times$ 0.219

3. 76.196
    $\times$ 0.072

4.   48.13 inches
    $\times$30.07

5.   0.5553 pounds
    $\times$ 13.63

6.   31.974 minutes
    $\times$ 3.65

7. 8.683 $\times$ 0.4039 _____

8. 490.21 $\times$ 3.72 _____

9. 65.2 $\times$ 0.00017 _____

10. 2.398 $\times$ 5.3 feet _____

11. 397 cu m $\times$ 0.1031 _____

12. 41.59 liters $\times$ 34.23 _____

13. Convert 6.2 feet to inches. _____

14. Convert 3.4 pounds to ounces. _____

15. In all liquid refrigerant lines, the pressure at the bottom of a vertical rise is greater than the pressure at the top of the rise. For R-134a, the pressure is about 0.5144 lb/sq in or less for every vertical foot of pipe that is used. The vertical rise between this condenser and evaporator is 7.5 feet. How much less is the pressure at the top of the rise than at the bottom? _____

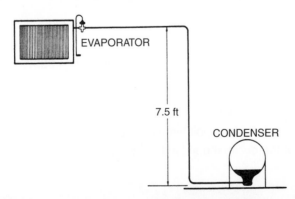

16. The rating stamped on an oil burner nozzle tells the number of gallons of oil sprayed each hour. One day the nozzle sprayed oil for 8.25 hours. How many gallons were sprayed during that day?

_____

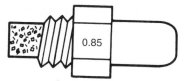

17. A ½-inch diameter fuel line runs from an oil tank to the burner. The line is made of ½-inch copper tubing weighing 0.344 pound per foot. What is the total weight of the tubing?

_____

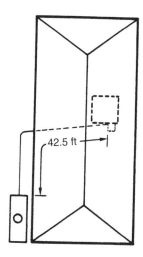

18. At a certain house, 147.3 gallons of *#2* fuel oil are unloaded. The fuel oil weighs 7.1 lb/gal. Find in pounds the total weight of the fuel oil unloaded.

_____

19. A mullion heater prevents condensation on the refrigerator cabinet between the two doors of the cabinet. One mullion heater has a value of 12.5 watts. One watt produces 3.415 Btu of heat. How much heat does the mullion heater produce?

_____

20. A technician's assistant earns $13.70 per hour. How much is the assistant paid for a job lasting 4.5 hours?

_____

21. When 1 pound of refrigerant R-134a vaporizes, 90.2 Btu of heat are removed from the surroundings. How many Btu of heat are removed when 9.4 pounds of R-134a vaporize?

_____

22. One watt of energy is equivalent to 3.41 Btu/h. An old storeroom is converted into a movie studio. Four 650-watt spotlights are added to the room. How much extra heat is produced by these spotlights when they are used?   _____

23. The manufacturer's manual for a fan motor states that the motor draws a starting current that is 6.2 times larger than its running current. If its running current is 2.147 amperes, find the expected current reading on an ammeter when starting the fan.   _____

24. At 86°F, the density of liquid refrigerant R-134a is 74.074 lb/cu ft. This means that at 86°F, 1 cu ft of liquid R-134a weighs 74.074 pounds. At 86°F, how much would 1.45 cubic feet of liquid R-134a weigh?   _____

25. At 5°F, the latent heat of vaporization (the heat needed to vaporize a liquid to a vapor) of R-124 is 70.7 Btu/lb. How many Btus of heat are needed to vaporize 18.6 pounds of R-124?   _____

Unit 14 DIVISION OF DECIMAL FRACTIONS

BASIC PRINCIPLES OF DIVISION OF DECIMAL FRACTIONS

*   Study division of denominate numbers in Section I of the Appendix.

There are a few differences between dividing whole numbers and dividing decimal numbers. The problems are still written in the same way. Remember that the general way to write a division problem is

$$\text{DIVISOR} \overline{)\text{DIVIDEND}}^{\text{QUOTIENT}}$$

Example: Divide 5.742 by 1.65.

$$1.65 \overline{)5.742}$$

But before dividing, move the decimal point in the divisor so that it is now a whole number. In this case, the decimal point would be moved 2 places to the right. Then move the decimal point the same number of places to the right in the dividend, and place the decimal point directly above that point in the quotient.

$$1\,65. \overline{)5\,74.2}^{\textbf{.}}$$

Now ignore the decimal point and divide as if the numbers were whole numbers. Be careful to place the number in the quotient in the correct place.

$$
\begin{array}{r}
3.4 \\
1\,65. \overline{)5\,74.2} \\
4\,95 \\
\hline
79\,2 \\
66\,0 \\
\hline
13\,2
\end{array}
$$

Now we can add zeros onto the end of the dividend to carry the division out further.

$$
\begin{array}{r}
3.48 \\
1\,65. \overline{)5\,74.20} \\
4\,95 \\
\hline
79\,2 \\
66\,0 \\
\hline
13\,20 \\
13\,20 \\
\hline
\end{array}
$$

There are times when the division will not come out evenly or when we do not need the answer to be more than a certain number of decimal places. The answer should be rounded off in these cases.

To round off, carry the division one place further than is asked for in the answer (hundredths if tenths were called for, thousandths if hundredths were called for, and so on). Look at the last number. If it is less then 5, drop the last number and you have your answer. If the number is 5 or larger, drop the last number, but raise the previous number by one to give you the answer. For example, 3.24 rounded off to the nearest tenth would be 3.2 (because 4 is less than 5); 151.348 rounded off to the nearest hundredth would be 151.35 (8 is larger than 5, so drop the 8 and make the 4 a 5).

Estimating can be very helpful when dividing decimal fractions, even when doing it longhand. Your answer should be close to what your estimate is.

To estimate this answer, the problem should be rounded off as

$$1.65 \overline{)5.742} \qquad \Rightarrow \qquad 2 \overline{)6}$$

The estimated answer is 3.

Dividing using the calculator, we enter 5.742 ÷ 1.65 = . This gives an answer of 3.48.

Remember: When dividing decimals longhand, move the decimal point to the right to make the divisor a whole number. Move the decimal point the same number of places to the right in the dividend and place it above that point in the quotient.

PRACTICAL PROBLEMS

Note: In problems 1–12, divide the decimal fractions.

1. $17 \overline{)59.84}$

2. $5.3 \overline{)343.175}$

3. $0.375 \overline{)0.15675}$

4. $8.09 \overline{)760.7836}$ cu in

5. $0.00261 \overline{)0.01963242}$ gallon
      (Round to the nearest thousandths.)

6. $6.17 \overline{)0.466708}$ meter

7.   2,073 ÷ 3.9 (Round the answer to the nearest tenth.)   _____

8.   31.71 ÷ 0.0755   _____

9.   2,844.686 ÷ 923   _____

10. 9.465712 cu yd ÷ 4.714   _____

11. 937.135 inches ÷ 28.1 inches   _____

12. 0.787826 liter ÷ 0.0013   _____

13. A technician is troubleshooting a problem in an electrical circuit. Seven identical air conditioners are running on one circuit. With all of them running, 12.845 amperes of current flow through the circuit. What should each air conditioner have as current running through its unit?   _____

14. A 50-foot coil of ⅜-inch diameter copper tubing weighs 9.9 pounds. What is the weight of 1 foot of the tubing?   _____

15. Air-conditioning units come in sizes of whole and half tons. A 1-ton air-conditioning unit will cool a typical 1,100 sq ft house in the southern part of the United States. What size unit would be needed to cool a 2,973 sq ft house in the southern part of the United States? (Round up to the next higher whole or half ton.)   _____

16. A stack of duct metal is 4 inches high. If there are 128 sheets in the stack, what is the thickness of each sheet?   _____

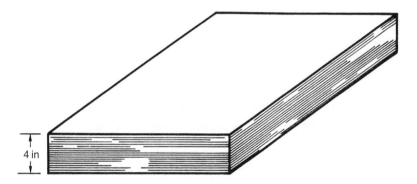

4 in

17. An oil burner ran a total of 4.5 hours in one day and used 7.425 gallons of fuel. How many gallons would be used if it ran only 1 hour?   _____

18. A technician is paid $84.51 for a job. The job takes 6.75 hours. What is the technician's hourly rate?   _____

Note: Solve problem 19 using the following information:

The rate at which heat flows through a substance is called *thermal conductivity.* Thermal conductivity of a substance is given as a U value. The U value has units of British thermal units, area in sq ft, time in hours, and temperature change in degrees Fahrenheit (Btu/sq ft • h • °F). The U value is related to the thermal resistance, or R value, by the formula

$$U = \frac{1}{R}$$

19.   The R value for the wall of a house is 3.51. What is the U value for the wall? Round the answer to the nearer ten-thousandth.   _____

20.   The *density* of a substance is the weight of the substance divided by its volume. At 5°F, the weight of 3.5 cu ft of liquid R-134a is 294.18 pounds. What is the density in lb/cu ft?   _____

21.   When a refrigerant vaporizes, it takes heat from its surroundings. The *latent heat of vaporization* is the amount of heat needed for the refrigerant to vaporize. When 38.5 pounds of R-124 vaporize at 5°F, the latent heat of vaporization is 2,721.95 Btu. What is the latent heat of vaporization for 1 pound of R-124 at 5°F?   _____

22.   In a certain air duct, pressure is being measured in inches of water. The duct is 26 feet long. Find in inches of water per foot the pressure loss in the duct.   _____

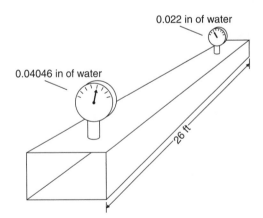

0.022 in of water

0.04046 in of water

26 ft

23. Find the wall thickness of this rigid polyvinylchloride (PVC) tube.   _____

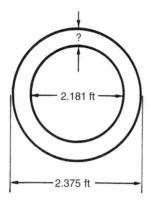

2.181 ft

2.375 ft

24. A 1-ton HVAC unit will remove up to 12,000 Btu/h. What size unit is needed
     for a house with a heat load of 28,576 Btu/h? (Round up to the next higher
     either half or whole number.)   _____

25. A circulating fan can supply 9,525.6 cu ft of air in 1 hour. What is the largest
     volume room that will have its air exchanged 4.5 times each hour?   _____

Unit 15 DECIMAL AND COMMON FRACTION EQUIVALENTS

BASIC PRINCIPLES OF DECIMAL AND COMMON FRACTION EQUIVALENTS

• Study denominate numbers in Section I of the Appendix.

Decimal fractions and common fractions are both types of fractions. Therefore, there should be some way of getting from one type of fraction to the other. Often the word *fraction* is left off when talking about decimal fractions and they are simply referred to as decimals.

Problems involving common fractions are best worked longhand. The main reason for this is that many common fractions do not have exact decimal fraction values. So when some common fractions are converted to decimal fractions on the calculator, the values the calculator uses are not exact, so the final answer is not exactly correct. In other words, you will get a slightly different answer using the calculator than by working the problem by hand. A second problem with using calculators is that often the answer must be expressed as a common fraction. So in order to work common fraction problems on the calculator, you must convert the common fraction to a decimal, then work the problem, and then convert the answer back to a common fraction.

Common fractions related to measurements using a ruler (8ths, 16ths, 32nds, and 64ths) all have exact decimal equivalents. Problems involving these numbers can be worked exactly using a calculator. To get the decimal fraction back to a common fraction, it is easiest to look up the equivalent in a table. The table in Section II of the Appendix contains the equivalents for fractions involving halves, quarters, eighths, and sixteenths.

Using a calculator to work problems with mixed numbers (whole numbers and fractions) requires special attention so that the numbers are correctly entered into the calculator. It can be done and done accurately, but special care must be taken. All in all, it may be easier to work problems with common fractions out longhand.

To convert from a decimal to a common fraction, take the number to the right of the decimal point and make that the numerator of the common fraction. The denominator is the value of the place of the last number of the decimal.

Example 1: The number 0.5 represents five-tenths. The fractional equivalent is ⁵⁄10, which like other fractions can be reduced. This one can be reduced to ½.

Another example is 0.431, or 431 thousandths, which has the common fractional equivalent of ⁴³¹⁄1000.

To convert from a common fraction to a decimal, divide the denominator into the numerator and carry the answer out as a decimal.

Example 2: The fraction ¼ can be converted to a decimal by dividing 1 by 4.

$$\begin{array}{r} 0.25 \\ 4\,\overline{)1.00} \\ \underline{8} \\ 20 \\ \underline{20} \end{array}$$

A problem that can occur when doing this process is that many common fractions do not have even decimal equivalents. So when the decimal starts to repeat itself, you may want to carry it out just a couple of decimal places unless otherwise instructed.

Therefore, the decimal equivalent of ⅓ is 0.33, and the equivalent of ⅑ is 0.11.

Remember: To change any fraction into a decimal, divide the denominator into the numerator. Many fractions do not have exact decimal equivalents. They have to be rounded off.

PRACTICAL PROBLEMS

Note: In problems 1–5, express each common fraction as a decimal fraction.

1. ³⁄16 _____

2. ⁵⁄64 _____

3. ⁷⁄8 _____

4. ⁸⁄25 _____

5. ⁷⁄40 _____

Note: In problems 6–10, express each common fraction as a decimal fraction. Round each answer to four decimal places.

6. ⅓ _____

7. ⁴⁄15 _____

8. $\frac{3}{13}$ _____

9. $\frac{1}{6}$ _____

10. $\frac{5}{11}$ _____

> **Note:** In problems 11–13, do the mathematical operations by first converting the fraction to a decimal number and then performing the indicated operation.

11. $2\frac{5}{8} \div 0.312$ _____

12. $4\frac{1}{4}$ inches $-$ 1.264 inches _____

13. 117.264 $\times$ $3\frac{3}{4}$ _____

> **Note:** In problems 14–16, round off the answers to three decimal places.

14. 21.42 ounces $+$ $2\frac{2}{9}$ ounces _____

15. $64\frac{7}{8}$ $\times$ 7.21 _____

16. $3\frac{1}{5}$ $+$ $7\frac{2}{7}$ $+$ 43.97 _____

17. Find in inches the inside diameter of this water pipe for an automatic ice maker. Express the answer as a decimal fraction. _____

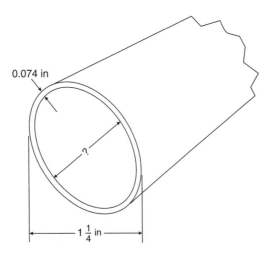

18. The gap between the electrodes of an oil burner gun is set at $\frac{1}{8}$ inch. Use has caused each electrode to wear 0.0155 inch. What is the new gap distance expressed as a decimal fraction? _____

19. The slight sideward motion of a shaft is called *end play.* The end play in the shaft of a rotor for an electric motor should not be more than $\frac{1}{32}$ inch. One motor has an end play of 0.0305 inch. Is this end play more than $\frac{1}{32}$ inch?   _____

20. A compressor has a volume of $1\frac{5}{8}$ cu in. On each stroke the compressor pumps 0.63 times its volume. What volume does the compressor pump on each stroke? Express the answer as a decimal fraction.   _____

21. What is the outside diameter of this piston ring, expressed as a decimal fraction?   _____

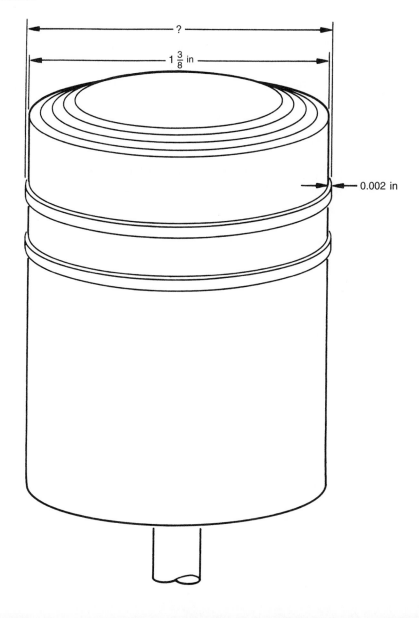

22. A major repair job on a refrigeration compressor took 16.5 hours to complete. Joe worked $\frac{1}{2}$ of that time, Susan worked $\frac{1}{3}$, while Al worked $\frac{1}{6}$ of the total. What time, expressed as a decimal, should be put on each worker's payroll sheet? Round each answer to the nearer tenth.   _____   _____   _____

Note: Solve problem 23 using the following information:

When an object goes on sale, the savings can be found by multiplying the portion taken off by the price of the object.

23. A heating supply dealer has taken $\frac{1}{5}$ off the price of an acetylene torch. If the price was $233.57, what is the savings for buying the torch today?   _____

24. The Energy Efficiency Rating (EER) of an air-conditioning unit is defined as the ratio of the heat removed in Btu/h divided by the energy used in watts. The EEI (Edison Electric Institute) names an air-conditioning unit as an Energy Star unit if it has an EER of 13 or greater. A $2\frac{1}{2}$-ton air-conditioning unit uses 2.41 kilowatts (2,410 watts) of energy. One ton is equivalent to 12,000 Btu/h. Can this unit be termed an Energy Star unit?   _____

25. A humidifier is designed to put 8.3 gallons of water into air flowing through an air duct every 24 hours when running continuously. How much water is put into the air when the system runs only $\frac{2}{5}$ of the time?   _____

Unit 16 COMBINED OPERATIONS WITH DECIMAL FRACTIONS

BASIC PRINCIPLES OF COMBINED OPERATIONS WITH DECIMAL FRACTIONS

- Review and apply the principles of addition, subtraction, multiplication, and division of decimal fractions to these problems.

- Read each problem. Decide which operation must be performed to solve the problem. Perform that operation.

- Some problems may require the application of more than one type of operation to solve the problem. As a result, it may take more than one step to find the answer.

PRACTICAL PROBLEMS

1.　　17.315
　　　　2.8047
　　＋ 0.7664

2.　　15.768
　　－ 3.503

3.　　　8.905
　　×0.0049

4.　8.41)38.0132

5.　4.437 centimeters + 0.0918 centimeter + 85.68 centimeters　　_____

6.　0.0057 gallon + 7.8994 gallons + 3.013 gallons + 11.652 gallons　　_____

7.　13.1 feet − 5.207 feet　　_____

8.　6.34922 sq yd − 3.7073 sq yd　　_____

9.　351.6 meters × 0.0545 meter　　_____

10.　0.0571 liter × 6.206 liters　　_____

11.　1.3383 cu in ÷ 0.527 cu in　　_____

12.　39 pounds ÷ 0.704 (Round the answer to the nearer tenth.)　　_____

Note: In problems 13 and 14, express each common fraction as a decimal fraction. Round to four decimal places when needed.

13. $\frac{3}{8}$ _____

14. $\frac{3}{20}$ _____

15.   What is the inside diameter of this pipe? _____

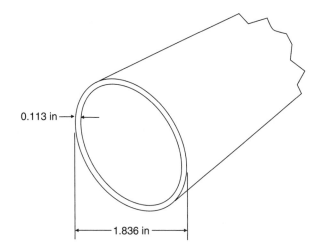

0.113 in →

1.836 in

16.   At 5°F, the latent heat of vaporization for R-124 is 70.7 Btu/lb. This means that when 1 pound of R-124 vaporizes at 5°F, 70.7 Btu of heat are removed from the surroundings. How many pounds of R-124 must vaporize at 5°F in order to remove 26,543.608 Btu of heat from the surroundings? _____

17.   A store replaced 60 of its 150-watt incandescent lightbulbs with new 45-watt compact fluorescent lightbulbs. The light output is the same; the difference in wattage is the difference in heat output by the bulbs. If 1 watt is equal to 3.41 Btus for each hour the light is on, how much less heat must be removed by the air conditioner each hour, due to changing the lightbulbs? _____

18.   By using a vertical U-tube, 2 feet of tubing can fit in each foot of well bore. A ground source heat pump requires 920 feet of heat exchanger tubing. How many wells 160 feet deep must be drilled to hold the tubing for this heat pump? (Round to the next higher whole number.) _____

19. This tube is to be used as a conduit for wires to a condenser unit. The conduit must pass through a wall. A $^{25}\!/_{32}$-inch drill is used to make the hole. Find the clearance around the tube. Express the answer as a decimal fraction.   _____

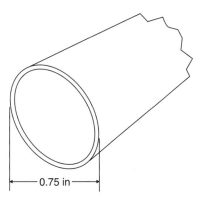

0.75 in

20. What is the total thickness of this refrigerator wall?   _____

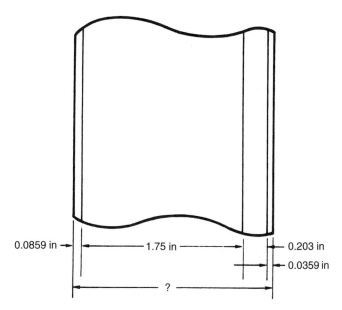

0.0859 in   1.75 in   0.203 in

0.0359 in

?

21. At 5°F, the latent heat of vaporization for R-134a is 90.2 Btu/lb and 70.7 Btu/lb for R-124. How much more heat is needed to vaporize 3 pounds of R-134a compared to R-124?   _____

22. Find in inches the thickness of the insulation in this refrigerator wall.

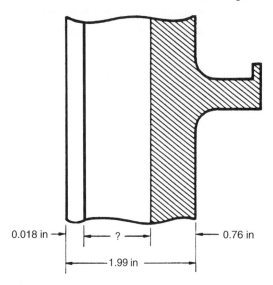

0.018 in →   ?   0.76 in

1.99 in

23. If a room that was used as a part of a house becomes office space and is air-conditioned rather than just heated like the house, the number of air changes per hour is increased. For the room in question, the number of air changes per hour becomes 2.3 times larger. If the old number of changes was 3.4 changes per hour and the room is 952.7 cu ft, what flow should the new ventilating system be able to handle in 1 hour?

24. An air duct is made from *26* gauge metal that is 0.0179 inch thick. The inside width of the duct is $8\frac{3}{16}$ inches. Find the outside width of the duct. Express the answer as a decimal fraction.

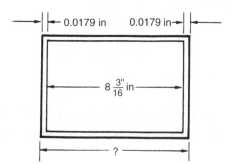

0.0179 in   0.0179 in→

$8\frac{3"}{16}$ in

?

25. Two technicians from the All Cool Company worked on a repair job. Janet earns $13.25 an hour and worked 10.5 hours. Bill, who earns $11.70 an hour, worked 9.75 hours. How much did the All Cool Company have to pay out for labor?

Ratio and Proportion

Unit 17 RATIO

BASIC PRINCIPLES OF RATIO

• Review denominate numbers in Section I of the Appendix.

Ratios are a way of comparing two numbers. When written as a mathematical expression, the ratio is written as two numbers separated by a colon (:). When written as a statement, a ratio is expressed as a ratio of one number **to** a second number. When written as a fraction, the first number of the ratio is the numerator of the fraction and the second number is the denominator.

In most cases, the ratio is a comparison of two whole numbers. So if the numbers being compared have fractions in them, an equivalent ratio is formed that has only whole numbers in it. For example, a ratio of $2\frac{1}{2}$ to 3 would be changed to a ratio of 5 to 6. (This is the same ratio and was found by multiplying both numbers by 2.) In the same way, a ratio of 2.5 to 3 would be changed to a ratio of 5 to 6.

A ratio can always be set up as a fraction. The ratio is usually stated as value A to value B. The quantity after the word **to** always becomes the denominator of the fraction.

There are special situations where the relationship between two numbers is known as an inverse ratio. The inverse ratio can be found by making the ratio and then dividing the number one by this ratio. Another way of getting the same result is to form the ratio and then find the inverse ratio by interchanging the numerator and the denominator.

There are two applications where ratios are often used. The first has to do with pulleys. When two pulleys with different diameters are connected by a belt, the revolutions per minute for each pulley are different. The ratio of the revolutions per minute is the inverse of the ratio of the pulley diameters.

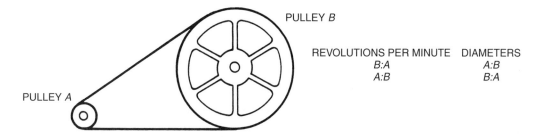

PULLEY B

REVOLUTIONS PER MINUTE	DIAMETERS
B:A	A:B
A:B	B:A

PULLEY A

The second application involves compression ratio. The compression ratio of a compressor is given as the pressure of the refrigerant at the discharge from the cylinder divided by the pressure of the refrigerant at the inlet into the cylinder. Pressure is measured in pounds per square inch absolute (psia).

$$\text{Compression ratio} = \frac{\text{Pressure at the discharge (in psia)}}{\text{Pressure at the inlet (in psia)}}$$

The ratio is then expressed as a decimal fraction.

This is one case where the ratio is often not two whole numbers, but is expressed as a decimal fraction (with the denominator being 1). For example, the compression ratio is 10.5 or 10.5:1.

PRACTICAL PROBLEMS

Note: In problems 1–4, express each ratio as a fraction in lowest terms.

1. 24:40   _____

2. 15:3   _____

3. $\frac{5}{8}$:$\frac{3}{8}$ _____

4. 12.2 to 7.4   _____

Note: In problems 5–7, find the inverse ratio of each ratio given. Express each answer as a fraction in lowest terms.

5. 7:3   _____

6. $\frac{6}{7}$:$\frac{2}{7}$ _____

7. 1.2 to 1.6   _____

Note: Use this diagram for problems 8–15.

A ├──────────────────────────────┤

B ├────────────────────┤

C ├─────────┤

D ├──────────────────┤

Measure each line to the nearer ⅛ inch. Using the measured lengths, find each ratio. Express each answer in lowest terms.

8.   A:B   _____

9.   A:C   _____

10. B:D   _____

11. C:D   _____

12. D:A   _____

13. C:B   _____

14. C:A   _____

15. D:B   _____

16. Refrigerant R-410A is a mixture of refrigerants R-32 and R-125. It takes 60 pounds of R-32 and 40 pounds of R-125 to make 100 pounds of R-410A. Find the ratio of R-32 to R-125.   _____

17. Find the ratio of the revolutions per minute for the fan pulley to the revolutions per minute for the motor pulley.   _____

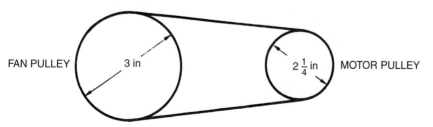

18. What is the ratio of the revolutions per minute for the motor to the revolutions per minute for the compressor?   _____

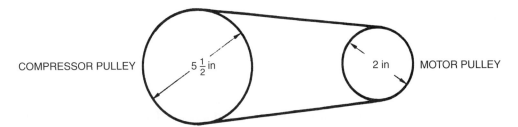

COMPRESSOR PULLEY $5\frac{1}{2}$ in   2 in   MOTOR PULLEY

19. The weight of 1 cu ft of #2 fuel oil is about 53.125 pounds. The weight of 1 cu ft of water is about 62.5 pounds. Find the ratio of the weight of the fuel to the weight of the water.   _____

20. Two new air-conditioning units are charged for the first time. To be fully charged, unit A needs 20 ounces of refrigerant. Unit B needs 14 ounces. Find the ratio of the amount of refrigerant used in unit A to the amount used in unit B.   _____

21. A compressor takes in refrigerant at a pressure of 80.34 psia. The discharge pressure of the refrigerant is 281.74 psia. What is the compression ratio of the compressor? Round the answer to the nearer tenth.   _____

22. A compressor takes in refrigerant R-134a at a pressure of 23.8 psia and compresses it to 122.8 psia. Find the compression ratio of the compressor. Round the answer to the nearer hundredth.   _____

23. In one minute, 90 cu ft of air flow through a duct into a room. The room contains 960 cu ft of space. What is the ratio of the flow of air into the room to the volume of the room?   _____

24. The intake of a compressor is 15 psig (pounds per square inch gauge reading). The exhaust has a pressure of 140 psig. Find the ratio of the exhaust pressure to the intake pressure.   _____

25. An installer compares the areas of a circular duct and a rectangular duct. The 6-inch by 4-inch rectangular duct has an area of 24 sq in. The 5-inch diameter circular duct has an area of about 20 sq in. What is the ratio of the area of the circular duct to the area of the rectangular duct?   _____

 *Unit 18* **PROPORTION**

BASIC PRINCIPLES OF PROPORTION

- Review denominate numbers in Section I of the Appendix.

Proportions are just two ratios set equal to each other. The following illustrates two ways of setting up a proportion (different letters are used as numbers):

$$A:B = C:D \text{ or } \frac{A}{B} = \frac{C}{D}$$

This can always be solved by cross multiplying:

$$A \times D = B \times C$$

Then, to find the unknown quantity, divide the two numbers being multiplied together by the number that is multiplied by the value you do not know. So, if A is the quantity that is not known, it can be found using:

$$A = \frac{B \times C}{D}$$

If B is the quantity that is unknown, it can be found using:

$$B = \frac{A \times D}{C}$$

Similar expressions could be written for C and D as the unknown.

Example: Complete the proportion: $4/10 = Y/75$

$$\frac{4}{10} = \frac{Y}{75}$$
$$4 \times 75 = 10 \times Y$$
$$10 \times Y = 4 \times 75$$
$$Y = \frac{4 \times 75}{10} = \frac{300}{10} = 30$$

A couple of applications that use proportion are:

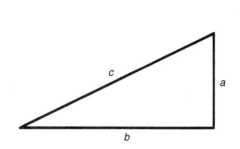
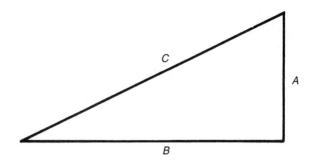

Corresponding sides of similar triangles (triangles that have equal angles) are in proportion. That is:

$$\frac{\text{side } a}{\text{side } A} = \frac{\text{side } b}{\text{side } B} \qquad \frac{\text{side } a}{\text{side } A} = \frac{\text{side } c}{\text{side } C} \qquad \frac{\text{side } b}{\text{side } B} = \frac{\text{side } c}{\text{side } C}$$

Another application involves pulleys. When dealing with pulleys connected by a belt, the proportion of pulley speeds to pulley diameters is set up as an inverse proportion. It is set up as:

$$\frac{\text{Diameter 1}}{\text{Diameter 2}} = \frac{\text{Speed pulley 2}}{\text{Speed pulley 1}}$$

PRACTICAL PROBLEMS

> **Note:**   In problems 1–8, find each unknown value.

1. $5/9 = x/27$ _____

2. $7/21 = 6/x$ _____

3. $2/5 = x/17$ _____

4. $13.5/6 = x/30$ _____

5. $2:9 = 6:?$ _____

6. $11:3 = ?:15$ _____

7. $3:5 = 7:?$ _____

8. $6.7:4 = x:30$ _____

9. The weight of 10 gallons of *#2* fuel oil is 71 pounds. What is the weight of 225 gallons?

   _____

10. The weight of 4 feet of 1-inch copper tubing is 2.62 pounds. Find the weight of 20 feet of the tubing.

    _____

11. A 5-foot section of 14" × 8" rectangular metal ducting weighs 18 pounds. What would be the weight of a 23-foot section of 14" × 8" rectangular duct?

    _____

12. A technician is able to tune up 3 oil burner systems in 7 hours. How many oil burner systems can the technician tune up in a 45-hour week?

    _____

13. Two triangles are similar. Triangle 1 has side a = 15 and side b = 8. Triangle 2 has side B = 30. How long is side A of triangle 2?

    _____

14. Find the length of the unknown side.

    _____

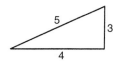 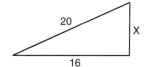

15. Measurements on this Y duct are taken from the center of the duct. Find dimension X.

    _____

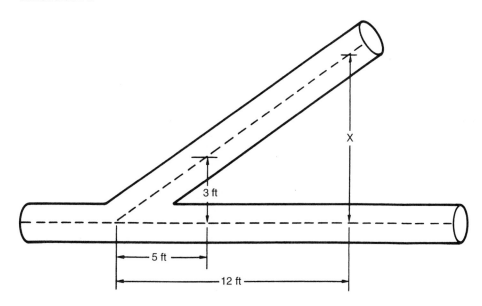

16. A fuel line from the tanks to the oil burner goes through the wall. Find
dimension Y.   _____

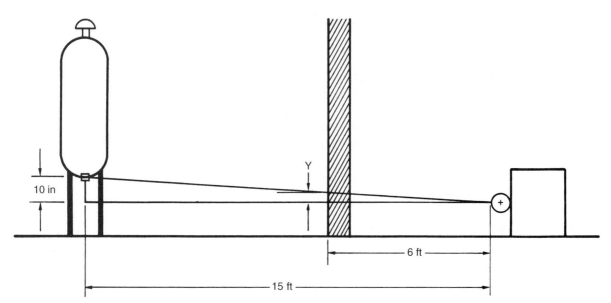

17. A compressor runs at a speed of 1,800 revolutions each minute. On each
revolution, 7.2 cu in of gas enter the compressor. What is the volume of gas
that enters the compressor each minute when it is running?   _____

18. If a 42,000-Btu wood-burning furnace can heat a 1,300 sq ft home, how large
a furnace is needed for a 900 sq ft cabin?   _____

19. For each 4 feet of duct, the pressure in the duct drops 0.003 inch of water. What is the pressure drop between the ends of this duct?

_____

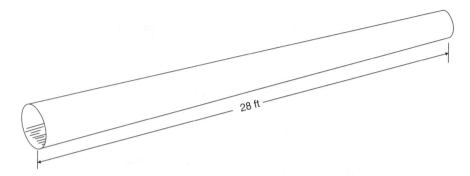

20. This motor pulley has a diameter of 2 inches and runs at 1,400 revolutions per minute (rpm). The fan pulley has a diameter of 3½ inches. At how many revolutions per minute does the fan pulley revolve?

_____

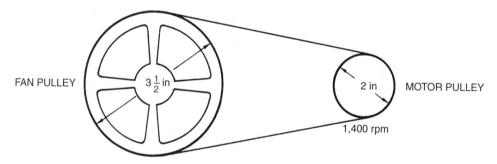

21. A compressor is run by a motor. If the compressor runs at 500 rpm, at how many revolutions per minute does the motor run?

_____

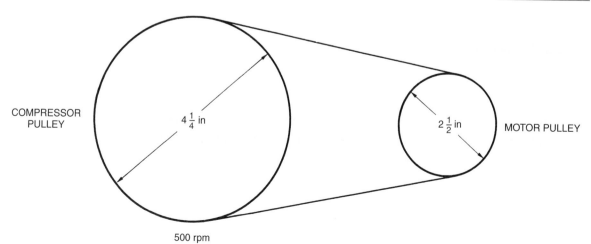

22.   This blueprint shows the layout for rectangular ducting for a home forced air heating system. The dimensions are to be measured along the center of the ducts. Find, in feet, each dimension.

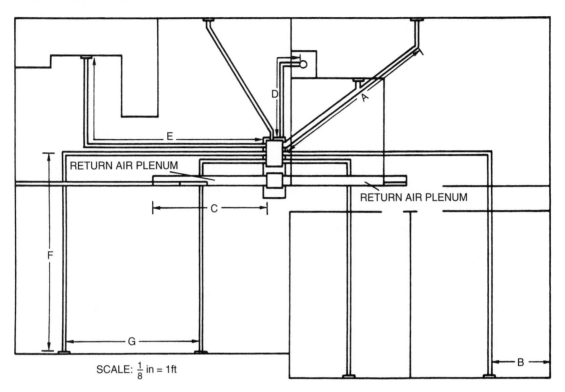

RETURN AIR PLENUM

RETURN AIR PLENUM

SCALE: $\frac{1}{8}$ in = 1ft

    a. Dimension A   _____

    b. Dimension B   _____

    c. Dimension C   _____

    d. Dimension D   _____

    e. Dimension E   _____

    f.   Dimension F   _____

    g. Dimension G   _____

23.   It has been determined that a 768 sq ft house would require a 5-ton air conditioner. For this same region, what size unit would a 1,080 sq ft house require? Round off to the next higher half-ton value.   _____

24. A 10,000-Btu window air conditioner requires 12 ounces of R-134a refrigerant to completely recharge the system. How much R-134a would be needed to refill a proportionally larger 12,000-Btu window air conditioner? _____

25. R-32 and R-125 are mixed to make R-410A. For every 60 pounds of R-32 used, 40 pounds of R-125 are needed. How many pounds of R-125 are needed when using 280 pounds of R-32? _____

Percent, Percentage, and Discount

Unit 19 PERCENT AND PERCENTAGE

BASIC PRINCIPLES OF PERCENT AND PERCENTAGE

* Review denominate numbers in Section I of the Appendix.

Percent is a method of writing decimals as whole numbers. The percent sign (%) takes the place of two decimal places. A percent is usually given with the percent sign. When doing a mathematical operation with the percent, it must first be changed to a decimal number. To change the number from a percent to a decimal, drop the percent sign and move the decimal point two places to the left; 35% becomes 0.35, while 50% becomes 0.5 (you drop zeros at the end of decimals). The decimal point is always moved two places to the left—so, 2% becomes 0.02 (a zero had to be added in front of the 2 because there were not two places to move the decimal point).

When writing a decimal as a number with a percent sign, just do the opposite of what was done above. Move the decimal point two places to the right and add the percent sign. Convert a common fraction to its decimal equivalent before writing it with a percent.

When trying to find the percent, form a ratio of $\dfrac{\text{part}}{\text{whole amount}}$ and find the decimal equivalent of this fraction. Then convert it to a number with the percent sign.

Almost all percentages can be written in the form:

Some % **of** a number (whole) **is** another number (part)

The statement can always be rewritten as a mathematical problem by putting a $\times$ sign in place of the word **of** and an $=$ sign in place of the word **is.** This means that the problem can always be set up as a ratio:

$$\% = \frac{\text{part}}{\text{whole}}$$

If the % is converted to a decimal, then this type of problem can be set up as a proportion

$$\frac{\%\ (\text{converted to a decimal})}{1} = \frac{\text{part}}{\text{whole}}$$

which can then be solved.

The values that are given in the problem can be identified with the help of the generalized form given above and then substituted into the proportion. The way the proportion is solved depends upon which number is not known.

Example 1:   Find 12% of 15. The statement can be thought of as 12% of 15 is what?

$$12\% = .12$$

$$\frac{.12}{1} = \frac{?}{15}$$

$$.12 \times 15 = ? \times 1 = ?$$

$$? = 1.8$$

Example 2:   6 is 10% of what number? This can be read as 10% of what number is 6?

$$10\% = .1$$

$$\frac{.1}{1} = \frac{6}{?}$$

$$? \times .1 = 6 \times 1 = 6$$

$$? = \frac{6}{.1}$$

$$? = 60$$

Example 3:   10 is what % of 40? This can be thought of as what % of 40 is 10?

$$\frac{?}{1} = \frac{10}{40} = .25$$

$$? = .25 \times 1 = .25$$

$$.25 = 25\%$$

PRACTICAL PROBLEMS

Note: In problems 1–4, express each percent as a decimal number.

1. 7.2% _____

2. 35$\frac{1}{2}$% _____

3. 83.7% _____

4. 125% _____

5. What is 5% of 83? _____

6. Find 27% of 115. _____

7. What percent of 175 is 250? Round the answer to the nearer whole percent. _____

8. The number 45 is what percent of 70? Round the answer to the nearer whole percent.

   _____

9. The number 24 is 15% of what number?

   _____

10. The number 125 is 70% of what number? Round the answer to the nearer hundredth.

    _____

11. An air-conditioning installer works part time and has a taxable income of $6,520. The state income tax is 7% of the taxable income. How much money does the installer pay in state taxes?

    _____

12. In one day, the blower fan of a forced air heating system runs for 9 hours. What percent of the day does the fan run? Round the answer to the nearer whole percent.

    _____

13. During a 40-hour workweek, a technician spends 15% of the time driving to and from various jobs. How many hours are spent driving?

    _____

14. In a heat pump, 15% of the heat is from the heat generated running the compressor. The rest is from the exchange of heat with the air or ground. How much heat is generated running the compressor of a 1½-ton heat pump? (1 ton is equivalent to 12,000 Btu/hr.)

    _____

15. A repair company borrows money to purchase new trucks. The interest paid on the loan is $1,440. This is 6% of the loan. How much money is borrowed?

    _____

16. A 250-gallon fuel oil tank has 100 gallons of fuel left in it. What percent of the tank is full?

    _____

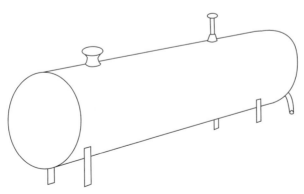

17. In charging a car air conditioner, 2.4% of a 15-pound cylinder of refrigerant R-134a is used. How many pounds of R-134a are used?

_____

18. In making a section of an air duct, 18 sq ft of sheet metal are used. This section is cut from a 24-sq ft sheet. What percent of the sheet is used?

_____

PORTION USED FOR DUCT

} 24-sq ft SHEET

19. A technician charges $60 for fixing a home heating system. This is 20% of the total bill. Find the amount of the total bill.

_____

20. An installer works for 6 hours and completes 48% of a job. How many hours are needed for the complete installation?

_____

21. In 1 hour, the air in a certain room is completely replaced 4 times. What percent of the air is replaced in 5 minutes? Round the answer to the nearer whole percent.

_____

22. An air-conditioning system is installed in a building under construction. The bill for installation is the cost of the parts plus overhead plus sales tax. The overhead is 75% of the cost of the parts. The parts cost $3,500. The sales tax is 4% of the cost of parts plus overhead.

   a. Find the cost of the overhead.

_____

   b. Find the sales tax.

_____

   c. Find the total amount of the bill.

_____

23. An air-conditioning contractor is installing a heat pump system in a private home. He has agreed to do the job for the cost of parts and labor plus a contractor's fee of 9% of the total. The heat pump costs $1,495, the electrical supplies $53, the ducting $81, and the other supplies $39. Three men work 16 hours each to install the system. They are paid $14 per hour. Sales tax is 6% of the subtotal of the cost for parts, labor, and contractor's fee.

a. What is the total cost for parts and labor?   _____

b. What is the contractor's fee?   _____

c. How much is the sales tax?   _____

d. What is the total bill to the homeowner?   _____

24. A heating supply store buys 6-inch metal flue pipe at $2.90 per 24-inch section. The pipe is marked up 40%. What is the selling price of one section?   _____

25. A heating contractor, in order to cover her expenses, sells parts with a 37% markup. If she bought 1/4-inch copper fuel line tubing for $0.53 a foot, what would be its selling price? Round to the nearer cent.   _____

Unit 20 DISCOUNTS

BASIC PRINCIPLES OF DISCOUNTS

- Review denominate numbers in Section I of the Appendix.

Discounts are percent problems with two exceptions. Discounts deal in money, so there will be decimals in both numbers. The second exception is that once the percent of a number has been taken, that value usually gets subtracted from the original price. This new price is often called the discounted price.

In most discount problems, the answer wanted is the new price. This is found by:

1. Multiplying the old price times the decimal equivalent of the percent. This gives the discount.

2. Subtracting the discount from the old price. This gives the new (lower) price.

There are times when customers are given double discounts. These may be given to preferred customers or cash-paying customers or for some other reason. The final price can be found using the following.

> First discount price = List price − (list price × first discount)
> Final price = First discounted price − (first discounted price × second discount)

There are also times when suppliers send bills to contractors with a notation such as 1% 10/Net 30. This means that there is a 1% discount if the bill is paid within 10 days. If not paid within 10 days, the full bill must be paid within 30 days.

PRACTICAL PROBLEMS

1. What is the discounted price of an item originally priced at $930 if the discount is 24%?

   _____

2. How much is saved on a $237 bill that has a discount of 34%?

   _____

3. A technician buys replacement parts needed to fix an air-conditioning system. The cost of the parts is $53.10. The distributor gives a 12% discount. What price does the technician pay? Round the answer to the nearer whole cent.

   _____

4. A repair company needs parts for an oil burner gun. A distributor will charge $173.40. Another repair company will sell the parts at a 6% discount. If bought from the other repair company, what is the cost of the parts? Round the answer to the nearer whole cent. _____

5. During a sale, insulation for heating ducts is purchased from a distributor. The insulation is valued at $675, and there is a 25% discount. Find the sale price of the insulation. _____

6. A new compressor for a refrigerator costs $141.40. An 8% discount is given. Find to the nearer whole cent the cost of the compressor. _____

7. A 5,500-Btu/hr room air-conditioning unit costs $382. If three or more units are purchased, a 12% discount is given. What will be the cost of four units? _____

8. A hot water heating system is installed in a house. The system needs 163 feet of ¾-inch copper tubing. The tubing lists for $1.49 per foot. A 17% discount is given. How much does the tubing cost? Round the answer to the nearer whole cent. _____

9. A residential heat pump system is installed. The bill for the system is $3,950. If the bill is paid within 10 days, an 11% discount is given. If the bill is paid after the 10 days and before 30 days, a 5% discount is given.

    a. What is the cost of the system if the bill is paid on the 8th day? _____

    b. What is the cost of the system if the bill is paid on the 21st day? _____

10. A supply company offers discounts to contractors. The amounts of the discounts are: electrical supplies, 12%; ducts, 9%; plumbing supplies, 11%; hardware, 7%. This is an air-conditioning contractor's bill before discounts.

THE COMPLETE SUPPLY COMPANY "We have what you need!"		
ITEM	LIST	DISCOUNTED PRICE
300 ft #10 wire	$173.00	
250 ft of 4 in × 8 in ducts	687.50	
25–20 ft lengths 3/4 in copper tubing (rigid)	604.75	
6 boxes 1 × 8 sheet metal screws	32.40	
	TOTAL	
Thank you!		

Find to the nearer cent each discounted price.

a. #10 wire

b. ducts   _____

c. copper tubing   _____

d. sheet metal screws   _____

e. What is the total amount of the bill after the discounts?   _____

                                                    _____

11. A department store has a freezer that does not work. The original price of the freezer is $319.50. It is sold at a 45% discount. A refrigeration technician buys it and spends $32.45 in fixing it. How much does the refrigerator finally cost the technician? Round the answer to the nearer whole cent.

                                                    _____

12. A repair company needs parts for a window air conditioner. The parts list for $273.75. A local supply company gives the repair company a discount of 8%. A distributor in a city 75 miles away gives a discount of 20%. The round-trip cost to the city is $40.50.

a. If the parts are bought from the local company, what is the cost?   _____

b. If the parts are bought from the distributor, what is the cost?   _____

c. Which place is less expensive?   _____

13. A shop needs 480 pounds of refrigerant R-134a. A supplier charges $0.93 per pound. If the refrigerant is ordered in 125-pound cylinders, a 13% discount is given. If ordered in 30-pound cylinders, a 9% discount is given.

   a. What is the cost of 750 pounds of R-134a if it is ordered in 125-pound cylinders? Round the answer to the nearer whole cent.

   _____

   b. What is the cost of 750 pounds of R-134a if it is ordered in 30-pound cylinders? Round the answer to the nearer whole cent.

   _____

14. A gas furnace is purchased for a new home. The list price is $469. The supplier gives a discount of 11% to all customers and then a second discount of 3% to selected contractors. Find to the nearer whole cent the final cost of the furnace to a selected contractor.

   _____

15. When purchasing a heat pump unit, an installer is given discounts of 12% and 2%. The unit is priced at $3,400. What is the final price of the pump?

   _____

16. A shop buys nuts, bolts, washers, and electrical staples from a distributor. The cost of these supplies is $123.72. The distributor gives the shop discounts of 9% and 6%. How much does the shop pay for the supplies? Round the answer to the nearer whole cent.

   _____

17. A bill for duct insulation and furnace filters is $237.15 with the notation 2% 10/Net 30. How much is saved by paying the bill within 10 days?

   _____

18. A contractor gets a bill for $545 for supplies purchased. The bill is notated 1.5% 10/Net 30. What must be paid if the bill is paid within 10 days?

   _____

19. A supplier sends a $1,387.55 bill to a contractor. The bill is notated 1% 10/Net 30. What will the contractor pay if the bill is paid within 10 days?

   _____

20. The motor for a commercial condensing unit needs to be replaced. The cost of a new one is $239. If the bad motor is turned in, a 33% discount is given. What is the price for the motor if the bad one is turned in?

   _____

21. The Toasty Home Heating Company charges $79.90 to tune up an oil burner. During the months of July and August, a 12% discount is given to customers. How much is saved by getting a tuneup in July or August?

   _____

22. A heating and air-conditioning technician is buying electrical supplies from the Big Charge Electrical Supply Company. If the bill is paid in cash, a 4% discount is given. How much does the technician save by paying cash if his bill is $317.25?

_____

23. The owner of the Old Reliable Plumbing and Heating Store has an electric heater for sale. Its current price is $93.95. If the owner gives a 41% discount, he will sell the heater for what he paid for it. What did the owner pay to buy the heater? Round the answer to the nearer whole cent.

_____

24. A supplier provides ½-inch elbows at a price of $1.15 less 8% less 5%. What is the price for the elbows? Round values to the nearer whole cent.

_____

25. A 3-ton air conditioner condensing unit lists for $1,200. The supplier sells it to an installer at 11% off the list price. The installer sells it to a homeowner at 3% off the list price.

a. What is the price the installer paid for the condensing unit?

_____

b. What is the price the homeowner paid?

_____

c. How much did the installer make on the deal?

_____

Direct Measure

SECTION 6

Unit 21 EQUIVALENT UNITS OF TEMPERATURE MEASURE

BASIC PRINCIPLES OF EQUIVALENT UNITS OF TEMPERATURE MEASURE

- Study and apply these principles of equivalent units of temperature measure.

Four of the most widely used temperature scales are the Rankine scale, the Fahrenheit scale, the Celsius scale, and the Kelvin scale.

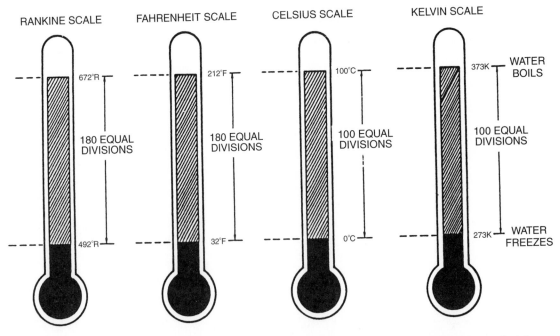

Most temperatures are measured using two different scales, the Fahrenheit and the Celsius scales. There may be times when you need to know what the temperature is on one of the scales when you have the value on the other scale.

There are two formulas to convert between these two temperature scales. To convert from the Celsius scale to the Fahrenheit scale, use

$$°F = \frac{9}{5}°C + 32$$

If the reading is in Fahrenheit degrees and you want to find the equivalent value on the Celsius scale, use

$$°C = \frac{5}{9}(°F - 32)$$

You must be careful when using these formulas to do the correct mathematical operations in the correct order. In the first formula where there are no parentheses, make sure that the C value is multiplied by $\frac{9}{5}$ and then the 32 is added. However, in the second formula, the F value has 32 subtracted from it first and then the result is multiplied by $\frac{5}{9}$ (notice that the F − 32 is in parentheses). You must do the operations in these orders.

Example 1: Convert 68°F to °C.

$$°C = \frac{5}{9}(°F - 32)$$

$$°C = \frac{5}{9}(68 - 32) = \frac{5}{9}(36) = 20$$

So 68°F is the same temperature as 20°C.

Example 2: Convert 15°C to °F.

$$°F = \frac{9}{5}°C + 32$$

$$°F = 1.8 \times °C + 32$$

$$°F = 1.8 \times 15 + 32 = 27 + 32 = 59$$

In this case, 15°C is the same as 59°F.

It is important to be able to convert temperatures, since they need to be converted to Rankine or Kelvin (absolute temperature scales) in order to correctly work the gas laws that you will be seeing later in the book.

It is easy to convert Celsius to Kelvin and Kelvin to Celsius. It is also easy to convert Fahrenheit to Rankine and Rankine to Fahrenheit. The formulas used to convert Celsius to Kelvin and Kelvin to Celsius are

$$°C = K - 273$$
$$K = °C + 273$$

Note: The unit for the Kelvin scale does not use the ° symbol and is called Kelvins instead of degrees Kelvin.

The formulas used to convert Fahrenheit to Rankine and Rankine to Fahrenheit are

$$°F = °R - 460$$

$$°R = °F + 460$$

When converting from Kelvin to Fahrenheit or Fahrenheit to Kelvin, make this a two-step problem. Find the Celsius temperature first. When converting from Rankine to Celsius or Celsius to Rankine, make this a two-step problem. Find the Fahrenheit temperature first.

Example 3:   Convert 113°F to K.

First, convert 113°F to °C.

$$°C = \frac{5}{9}(°F - 32)$$

$$°C = \frac{5}{9}(113 - 32) = \frac{5}{9}(81) = 45$$

Second, convert 45°C to K.

$$K = °C + 273$$

$$K = 45 + 273 = 318$$

There are times when it is not important what the actual temperature is, but what the temperature difference is. You may have to convert the temperature difference from one temperature scale to another. To express a temperature difference on the Fahrenheit scale as a temperature difference on the Celsius scale, use the equivalence:

$$\text{difference in } °C = \frac{5}{9} \times \text{difference in } °F$$

To express a temperature difference on the Celsius scale as a temperature difference on the Fahrenheit scale, use the equivalence:

$$\text{difference in } °F = \frac{9}{5} \times \text{difference in } °C$$

Remember: When converting from Fahrenheit to Celsius, be sure to subtract first, then multiply.

When converting from Celsius to Fahrenheit, be sure to multiply first, then add.

When converting temperature differences from Fahrenheit to Celsius, multiply by $\frac{5}{9}$.

When converting temperature differences from Celsius to Fahrenheit, multiply by $\frac{9}{5}$.

PRACTICAL PROBLEMS

Note: In problems 1–3, express each Fahrenheit scale temperature as a Celsius scale temperature. Round to the nearer tenth when necessary.

1. 77°F   _____

2. 657°F   _____

3. 172°F   _____

Note: In problems 4–6, express each Celsius scale temperature as a Fahrenheit scale temperature.

4. 85°C   _____

5. 62°C   _____

6. 228°C   _____

Note: In problems 7 and 8, express each Celsius scale temperature as a Kelvin scale temperature.

7. 120°C   _____

8. 35°C   _____

Note: In problems 9 and 10, express each Fahrenheit scale temperature as a Rankine scale temperature.

9. 85°F   _____

10. 5°F   _____

11. Express 25°C in °R.   _____

12. Express 300K in °F.   _____

13. Convert 501°R to °C.   _____

14. Express 41°F in Kelvins.   _____

15. The temperature reading on this outdoor thermometer is in degrees Fahrenheit. What is the temperature reading in degrees Celsius?   _____

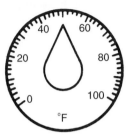

16. The water in a hot water heating system is heated to 76°C. What is this temperature reading on a Fahrenheit thermometer?   _____

17. This thermostat setting is in degrees Fahrenheit. What is the equivalent thermostat setting on the Celsius scale?   _____

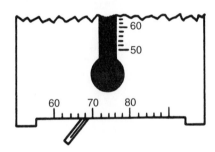

18. The temperature at which a fuel begins to burn is called the *ignition temperature*. The ignition temperature for *#2* fuel oil is 700°F. Find the ignition temperature in degrees Celsius. Round the answer to the nearer tenth.   _____

19. The temperature of the returned air of this warm air furnace is 65°F. The heated air is at a temperature of 140°F.

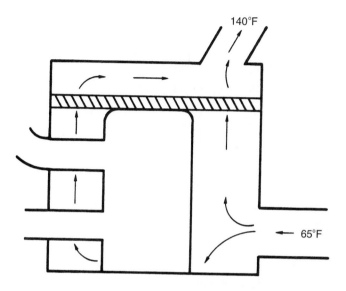

140°F

65°F

a. What is the temperature to the nearer tenth of the returned air expressed in degrees Celsius?

b. What is the temperature of the heated air expressed in degrees Celsius?

20. An air conditioner turns on when the air reaches a temperature of 80°F. The unit turns off when the air reaches 78°F.

a. At what Celsius scale temperature to the nearer tenth will the unit turn on?

b. At what Celsius scale temperature to the nearer tenth will the unit turn off?

21. The temperatures on this indoor-outdoor thermometer are in degrees Celsius.

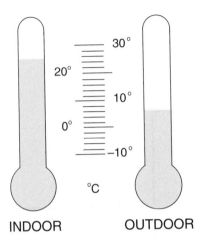

INDOOR   OUTDOOR

   a. What is the inside temperature in degrees Fahrenheit?   _____

   b. What is the outside temperature in degrees Fahrenheit?   _____

22. As a refrigerant travels through the cooling system of a refrigerator, the temperature of the refrigerant changes.

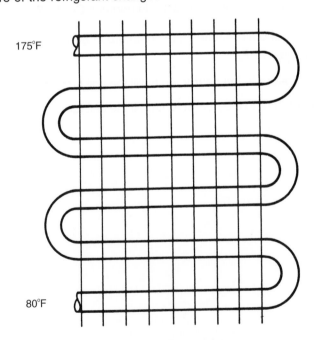

   a. Find in degrees Fahrenheit the temperature difference.   _____

   b. Find in degrees Celsius the temperature difference. Round the answer to the nearer tenth.   _____

23. The temperature difference between the floor and the ceiling of a room is 8°F. Express this difference in degrees Celsius. Round the answer to the nearer tenth.   _____

24. On a cold day, the temperature difference between the inside and the outside of a certain house is 25°C. Express this value on the Fahrenheit scale.   _____

25. A hydronic heating system is designed to use 180°F water leaving the furnace and returns the water to the furnace at 168°F.

   a. Find in degrees Celsius the temperature of the water leaving the furnace.   _____

   b. Find in degrees Celsius the temperature of the water returning to the furnace.   _____

Unit 22   ANGULAR MEASURE

BASIC PRINCIPLES OF ANGULAR MEASURE

• Study and apply the principles of angular measure.

This unit, unlike most of the others, has almost no calculations in it. It primarily involves simply measuring angles. Angles are measured with a protractor. If any calculations do have to be made, remember that a full circle has 360° in it.

A protractor is a half circle (semicircle) with degree markings on it. Although protractors come in different sizes and markings, all protractors do have some common points. As a rule, all protractors have degree markings for each degree. Sometimes the markings go from 0° to 180° in each direction, while others go from 0° to 90° to 0° again.

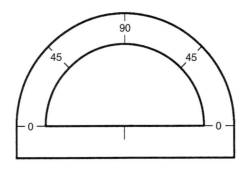

An angle is made when two straight lines meet. The two lines are known as sides, or legs, of the angle and they meet at the apex, or vertex, of the angle.

The size of an angle is measured by placing the center point of the protractor (the mark on the line between the two 0s) on the apex of the angle. Rotate the protractor so that the line to one of the 0s lies right on top of one of the sides of the angle while keeping the center point of the protractor on the apex of the angle. Next simply read off the angle where the other side of the angle crosses the protractor. If the angle is greater than 90°, care must be taken to read the correct scale or to count the number of degrees greater than 90. Many times the lengths of the sides of the angle are small. These sides can be extended before measuring the angle.

If the angle is greater than 180°, measure the smaller angle and then subtract that angle from 360°.

Remember: To measure an angle using a protractor, always place the center mark of the flat edge at the apex or vertex of the angle (the point where the two lines meet). Rotate the protractor around that point so that one section of the flat edge is aligned with one of the lines (sides) of the angle. The size of the angle is read where the other side of the angle crosses the curved section of the protractor. If the angle is larger than a straight line (180°), measure the smaller angle and subtract it from 360°.

PRACTICAL PROBLEMS

Measure each angle to the nearest degree.

1.

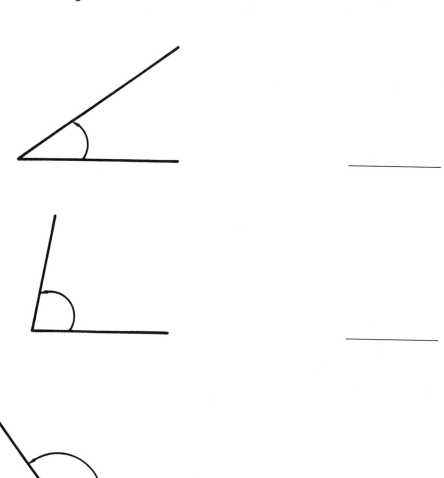

2.

3.

_____

4.

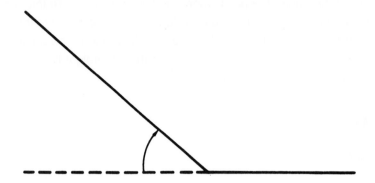

_____

5.

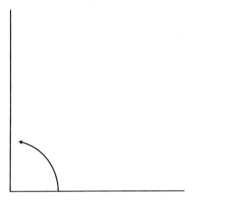

_____

6.

_____

7.

_____

8.

A. _____

B. _____

C. _____

9.

A. _____

B. _____

10.

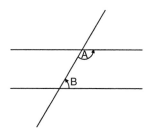

A. _____

B. _____

11.

A. _____

B. _____

12.

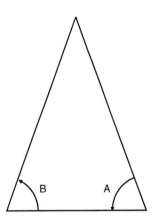

A. _____

B. _____

13.

A. _____

B. _____

C. _____

14.

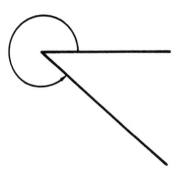

_____

15.

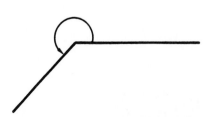

_____

16. How many degrees are in the angle formed by the Y in this duct? _____

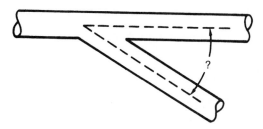

17. What is the angle made by the tip of this flaring cone? _____

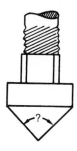

18. Find the number of degrees in the angle formed by the two parts of this A-type evaporator coil. _____

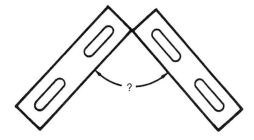

19. The control shown is for a window air conditioner.

    a. Through what angle must this dial be turned to go from Off to Low Cool?   _____

    b. Through what angle must this dial be turned to go from Off to High Cool?   _____

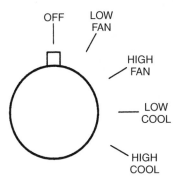

20. Find the angle through which this conduit has been bent.   _____

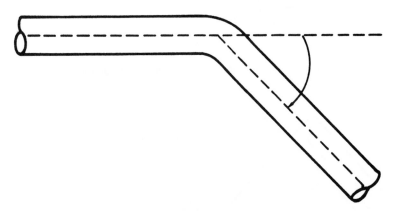

21. This damper is in the chimney flue of an oil burner. Through how many degrees is it turned from its closed position?   _____

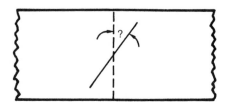

22.   What is the angle between the cylinders of this compressor?   _____

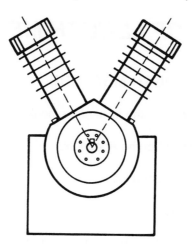

23.   A bolt is turned from 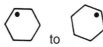 to ⬡ when tightening down on it.

Through what angle was the bolt turned?   _____

24.   A solar-heating collection panel needs to be oriented at 27° down from the
vertical for maximum winter performance. The roof of the house has an angle
of 30° with the horizontal. What angle will the brackets have that attach the
back of the collection panel to the roof?   _____

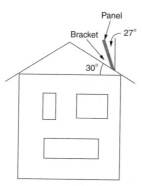

25.   When a hexagonal head bolt is turned through 60°, it looks the same as when
it started. Doing that is sometimes referred to as turning the bolt one flat (the
flat sides line up again). When tightening down on a flange to stop a leak at
a cooling system heat exchanger, one bolt was turned 2½ flats. Through what
angle was the bolt turned?   _____

Unit 23 UNITS OF LENGTH MEASURE

BASIC PRINCIPLES OF UNITS OF LENGTH MEASURE

- Review the Introduction to Measurement in the front of the book.

- Review denominate numbers in Section I of the Appendix.

- Review equivalents in Section II of the Appendix.

- Study the tables of units of length measure.

Often when solving problems involving units, all of the units must be the same. There are many times when the units in the problem are different, so they must be converted. There is a method that can always be used to convert from one unit to another correctly.

Suppose that we want to convert from 24 inches to feet. We all know that there are 12 inches to 1 foot, so 12 is the conversion factor; but do we multiply or divide by the 12? The following method will tell you which to do each time.

First, write

24 inches

Next to this write a fraction containing the units you are using so that the units you are starting with will get canceled. In this case

$$24 \text{ inches} \left(\frac{\text{feet}}{\text{inches}} \right)$$

The inches will cancel out of the fractions, leaving feet as the final unit. Now fill in the fraction with numbers so that the numerator and denominator are equivalent. We know that 1 foot is equal to 12 inches, so

$$24 \text{ inches} \left(\frac{1 \text{ foot}}{12 \text{ inches}} \right)$$

This is then solved just like a fraction problem. So the answer becomes

$$24 \text{ inches} \left(\frac{1 \text{ foot}}{12 \text{ inches}} \right) = 2 \text{ feet}$$

Getting to the unit that you want will often involve more than one conversion, since you do not always know the direct conversion factor. The conversion can still be done; however, it may take two or more fractions.

Example 1: Convert 15,840 inches to miles. We do not know the number of inches in a mile; however, we do know the number of inches in a foot and the number of feet in a mile, so we can make two conversions instead of one.

$$15,840 \text{ inches} \left(\frac{\text{feet}}{\text{inches}}\right) \left(\frac{\text{miles}}{\text{feet}}\right)$$

Once the units are straight, fill in the numbers. Remember that each fraction must be an equivalent, regardless of other fractions or numbers around.

$$15,840 \text{ inches} \left(\frac{1 \text{ foot}}{12 \text{ inches}}\right) \left(\frac{1 \text{ mile}}{5,280 \text{ feet}}\right) = 0.25 \text{ miles or } \frac{1}{4} \text{ mile}$$

There are times when numerators other than one get multiplied together. It depends upon the conversion that must be made.

Tables of some unit equivalents are included below.

ENGLISH LENGTH MEASURE		
1 foot (ft)	=	12 inches (in)
1 yard (yd)	=	3 feet (ft)
1 mile (mi)	=	1,760 yards (yd)
1 mile (mi)	=	5,280 feet (ft)

METRIC LENGTH MEASURE		
10 millimeters (mm)	=	1 centimeter (cm)
100 centimeters (cm)	=	1 meter (m)
1,000 meters (m)	=	1 kilometer (km)

To convert units correctly time after time, do the same process each time. An easy way to do this is to treat the problem as the original number and unit being multiplied by a fraction. The denominator of the fraction contains the original unit. The numerator has the unit the answer should have. The numbers that go with these units usually come from a chart that has these units as equivalents. Setting the problem up this way tells you whether to multiply or divide to get the answer.

Example 2: 3 feet = ? inches

$$3 \text{ feet} \times \frac{\text{inches}}{\text{feet}}$$

$$3 \text{ feet} \times \frac{12 \text{ inches}}{1 \text{ foot}}$$

$$3 \text{ feet} \times \frac{12 \text{ inches}}{1 \text{ foot}} = 36 \text{ inches}$$

Another way of looking at these problems is as follows.

The conversion process can be set up as a proportion. One of the ratios is the equivalence from the tables. The problem is then solved like a proportion problem. Make sure the units are in the same order in both ratios.

Example 3: Let us use the first example from this unit again. Convert 24 inches to feet.

$$24 \text{ inches} = ? \text{ feet}$$

$$\frac{24 \text{ inches}}{? \text{ feet}} = \frac{12 \text{ inches}}{1 \text{ foot}}$$

$$? \text{ feet} \times 12 \text{ inches} = 1 \text{ foot} \times 24 \text{ inches}$$

$$? \text{ feet} = \frac{1 \text{ foot} \times 24 \text{ inches}}{12 \text{ inches}} = 2 \text{ feet}$$

This is essentially what is being done with the fractions in the previous method of solving the problem. This method becomes a little difficult when there is more than one equivalence that must be used. The fractions work very well in that case; however, either system will work.

PRACTICAL PROBLEMS

Note: In problems 1–4, express each measurement in inches.

1.   5 feet   _____

2.   4½ feet   _____

3.   7⅙ feet   _____

4.   3.4 feet   _____

Note: In problems 5–8, express each measurement in centimeters.

5. 1.75 meters   _____

6. 3 meters   _____

7. 0.07 meter   _____

8. 2.505 meters   _____

Note: In problems 9–12, express each measurement in feet.

9. 60 inches   _____

10. 108 inches   _____

11. 50 inches   _____

12. 42 inches   _____

Note: In problems 13–16, express each measurement in meters.

13. 465 centimeters   _____

14. 29 centimeters   _____

15. 550 centimeters   _____

16. 0.9 centimeter   _____

17. Express 2 feet 7 inches as inches.   _____

18. Express 6 feet 10 inches as feet.   _____

19. Express 3 feet 9 inches as inches.   _____

20. A house is using radiant heat in the main-floor living room and entranceway. An electric cable is installed in a floor to be covered with mortar to create a radiant floor heating system. The cable is attached to pins that are installed 6 inches apart from one another. The cable is laid out as shown. Each end of the cable has a 4-foot length from the last pin to connect to the power supply. Find the length of cable in feet needed for this project.   _____

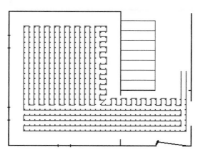

21.   Express the dimensions of each room in feet.

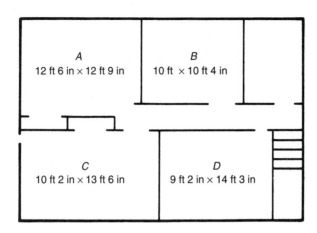

a. Room *A*   _____

b. Room *B*   _____

c. Room *C*   _____

d. Room *D*   _____

22.   The studs in a wall are spaced 1 foot 4 inches apart (center to center). The holes in an electric baseboard heater need to be how many inches apart so that each screw goes into a stud?   _____

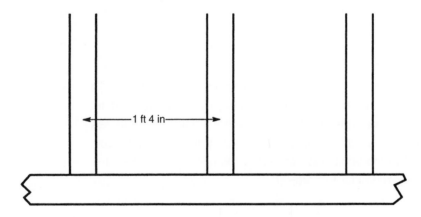

23.   A window is 2 feet 7 inches across. What is the largest width air conditioner
      in inches that will fit in that window?   _____

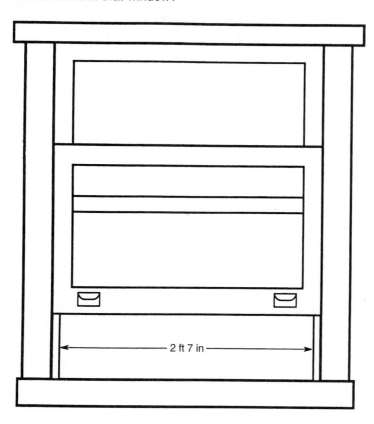

2 ft 7 in

24.   A strap to support a round duct is 1.27 meters long. Find the length of the
      strap in centimeters.   _____

25.   A domestic heat pump system has the condenser coils and evaporator coils
      separated by 10.46 meters. What would be the length of the hose, expressed
      in centimeters, connecting these two coils?   _____

Unit 24 EQUIVALENT UNITS OF LENGTH MEASURE

BASIC PRINCIPLES OF EQUIVALENT UNITS OF LENGTH MEASURE

- Review the Introduction to Measurement in the front of the book.

- Review denominate numbers in Section I of the Appendix.

- Review equivalents in Section II of the Appendix.

- Study the tables of equivalent units of length measure.

The United States is one of a very few countries in the world using the English unit system. A lot of other countries use the metric system. Items made in those countries are dimensioned in metric units, and many of their instructions include metric measurements. It is important to be able to quickly and correctly convert from one unit system to the other.

In Unit 23, methods were presented to convert between units of measure. These methods work whether in the English unit system, the metric unit system, or both systems. In the case of converting between unit systems, one of the fractions has to be the equivalent between English and metric units. The method is the same as before. As a matter of fact, this method will work when converting units of length, weight, area, volume, or any other type of unit that you want. This one method will do it all!

A table of equivalences between English and metric units is given below.

1 millimeter (mm)	=	0.03937 inch (in)
1 centimeter (cm)	=	0.3937 inch (in)
1 meter (m)	=	39.37 inches (in)

1 inch (in)	=	25.4 millimeters (mm)
1 inch (in)	=	2.54 centimeters (cm)
1 inch (in)	=	0.0254 meter (m)

1 foot (ft)	=	0.3048 meter (m)
1 yard (yd)	=	0.9144 meter (m)
1 mile (mi)	=	1.609 kilometers (km)
1 meter (m)	=	3.28084 feet (ft)
1 meter (m)	=	1.09361 yards (yd)
1 kilometer (km)	=	0.62137 mile (mi)

PRACTICAL PROBLEMS

Note: For problems 1–6, round to the nearer hundredth when needed.

1.   Express 9 inches as centimeters.   _____

2.   Express 2 feet as centimeters.   _____

3.   Express 1 foot 7 inches as centimeters.   _____

4.   Express 3 feet as meters.   _____

5.   Express 7 feet as meters.   _____

6.   Express 5 feet 8 inches as meters.   _____

   Note: For problems 7–12, round to the nearer thousandth when needed.

7.   Express 15 centimeters as inches.   _____

8.   Express 2 meters as inches.   _____

9.   Express 86 centimeters as feet and inches.   _____

10.   Express 4 meters as feet.   _____

11.   Express 7 meters as feet and inches.   _____

12.   Express 6.3 meters as feet and inches.   _____

13.   Find, in centimeters, the diameter of this duct.   _____

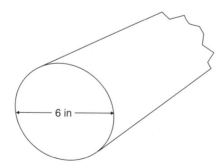

6 in

14.   The connecting rod in a compressor for an air-conditioning system is 3.5
   inches long. How long is the rod in centimeters?   _____

15.   The dimensions of this room are in meters.

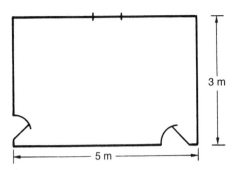

a. Find the width of the room in feet and inches. Round the inches to the nearer tenth.

_____

b. Find the length of the room in feet and inches. Round the inches to the nearer tenth.

_____

16.   An anemometer measures the velocity of the air from a grill. The anemometer reading is 18 ft/m. What is the reading in meters per minute?

_____

17.   The screws on this supply plenum are equally spaced. The center-to-center distance between screws is 25 centimeters. What is the distance in inches?

_____

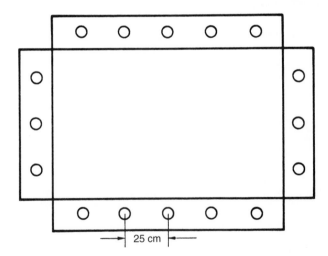

18.   A section of tubing for a baseboard hot water heating system must be replaced. The section is 2 feet 8 inches long. The replacement is measured in centimeters. What is the length of the tubing in centimeters?

_____

19. To get power to a compressor for an air conditioner, 11 feet of wire are
     needed. A roll of wire with 15 meters of wire on it is used. How many meters
     of wire are left?

     _____

20. A refrigerator door needs a gasket 3.48 meters long. The roll that the gasket
     is to be taken from contains 25 feet. How many feet of gasket are left? Round
     the answer to the nearer tenth foot.

     _____

21. The dimensions of this duct are in feet.

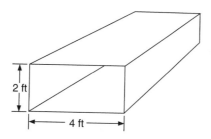

     a. Find in centimeters the width of the duct.

     _____

     b. Find in centimeters the height of the duct.

     _____

22. Flue pipe connects an oil furnace with the chimney of a house. The flue is 4
     feet long and is inclined so that it rises 6 inches from the furnace to the
     chimney.

     a. What is the length of the flue in meters?

     _____

     b. What is the rise of the flue in centimeters?

     _____

23.   A forced air system uses a filter 20 inches in length, as shown.

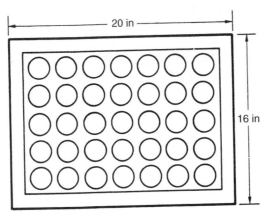

a. Find in centimeters the length of the filter.   _____

b. Find in centimeters the width of the filter.   _____

24.   The dimensions of this gas furnace are in inches. Find the dimensions in meters, and round each answer to the nearer tenth meter.

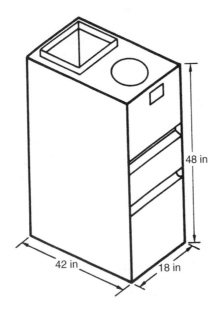

a. What is the height of the furnace?   _____

b. What is the width of the furnace?   _____

c. What is the length of the furnace?   _____

25.   A condensing coil has 12 fins per inch. How many fins per centimeter does the coil have? Round the answer to the nearer hundredth.   _____

Unit 25 LENGTH MEASURE

BASIC PRINCIPLES OF LENGTH MEASURE

- Review the Introduction to Measurement in the front of the book.

- Review denominate numbers in Section I of the Appendix.

- Review the tables of length measure in Section II of the Appendix.

- Use formulas for perimeters found in Section III of the Appendix.

Adding or subtracting lengths requires that all of the lengths be the same units. An exception to this might be when adding lengths in feet and inches: these could be added by converting all the measurements to feet or all to inches and then doing the addition; a second way to find the sum would be by adding all of the inches first. If the total is larger than 12 inches, convert to feet and inches; then add this feet value to the other feet values.

Example: Add 3 feet 5 inches, 4 feet 7 inches, and 8 feet 10 inches.

> 3 feet   5 inches
> 4 feet   7 inches
> <u>8 feet 10 inches</u>

First, add 5 inches + 7 inches + 10 inches = 22 inches. Next, 3 feet + 4 feet + 8 feet = 15 feet. Now 22 inches is larger than 12 inches, so convert it: 22 inches = 1 foot 10 inches. Then add this 1 foot to the 15 feet to get 16 feet. The final problem and answer become:

> 3 feet   5 inches
> 4 feet   7 inches
> <u>8 feet 10 inches</u>
> 15 feet 22 inches = 16 feet 10 inches

When subtracting, you can use a similar procedure by borrowing 1 foot and converting it to 12 inches before subtracting, if necessary.

Most of the time, the dimensions will all be in one unit and these types of conversions will not have to be done. Make sure that all of the measurements that are being combined have the same units (with the exception given above).

The distance around any figure is called the *perimeter*. The perimeter (*P*) of a rectangle is:

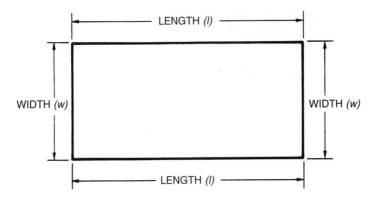

$$\text{perimeter} = \text{length} + \text{width} + \text{length} + \text{width}$$

or

$$P = 2l + 2w$$

The perimeter (*P*) of a square is:

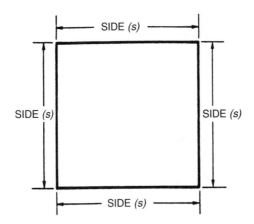

$$\text{perimeter} = \text{side} + \text{side} + \text{side} + \text{side}$$

or

$$P = 4s$$

The perimeter of a circle is called the *circumference*. The circumference (*C*) is:

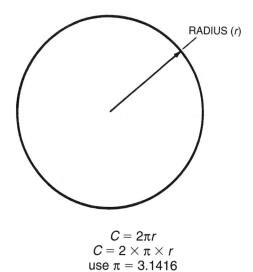

RADIUS (*r*)

$$C = 2\pi r$$
$$C = 2 \times \pi \times r$$
$$\text{use } \pi = 3.1416$$

There are times when the diameter (*D*) of the circle is given. The radius (*r*) is found by dividing the diameter by 2.

$$\text{Radius} = \frac{\text{Diameter}}{2}$$

$$r = \frac{D}{2}$$

$$D = 2r$$

DIAMETER (*D*)

PRACTICAL PROBLEMS

1. The duct shown must be wrapped with insulation. The width of the insulation is designed to wrap completely around the duct. Find in feet and inches the length of insulation needed to complete the job.   _____

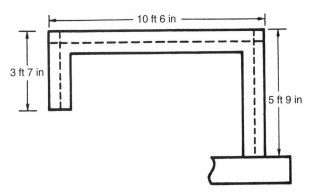

2. A piece of polyvinylchloride (PVC) drain pipe is used in an air-conditioning unit. The pipe is 243 centimeters long. It is cut from a pipe that is 5 meters in length. How much pipe is left?   _____

3. A flue from an oil furnace to the chimney is made from 9 pieces of pipe. After being fitted together, each piece of pipe is 1 foot 10½ inches long. What is the total length of the flue?   _____

4. Find in meters the length of straight duct used for this duct. (The corner pieces are separate parts and not included in the total.)

_____

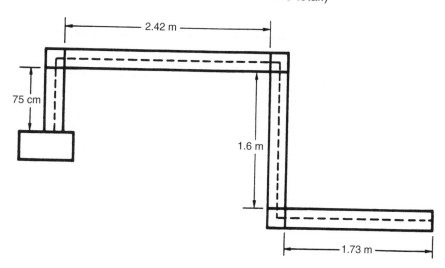

5. In repairing a condenser coil, a piece of tube 65 centimeters long is used.

    a. What is the length of tube needed for 16 pieces? Express the answer in centimeters.

    _____

    b. How many meters long are the 16 pieces of tube?

    _____

6. Support straps for a duct are each 1 foot 2 inches long. How many support straps can be cut from a 40-foot piece of metal?

    _____

7. The control wire from a thermostat to an oil burner control switch runs the following distances: 1½ inches, 5 feet 4½ inches, 3 inches, 5½ inches, 12 feet 6 inches, 7 feet 2¼ inches, 8 inches. Find the total length of the control wire.

    _____

8. When installing an air conditioner in a car, hose with a diameter of 2.5 centimeters is used. Each end of each hose used must overlap 2 centimeters. The lengths needed, without overlap, are 3.24 meters and 3.47 meters. What is the total length of hose the installer must cut?

    _____

9. The openings in this grill are equally spaced.

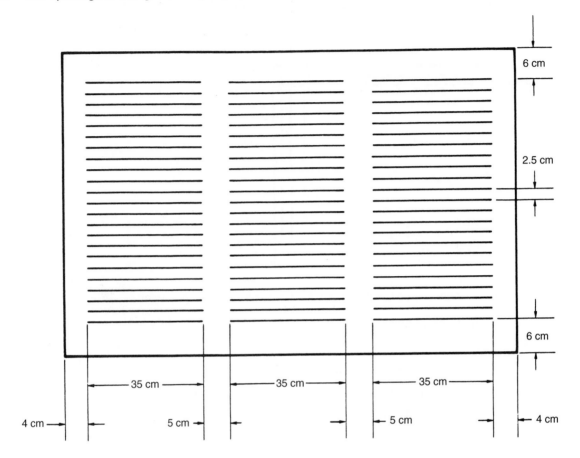

a. Find in meters the width of the grill.   _____

b. Find in meters the height of the grill.   _____

10. This electrical conduit is made from ⅝-inch diameter tubing joined to ⅝-inch diameter elbows. What is the total length of straight tubing used for the conduit? Express the answer in feet and inches.

_____

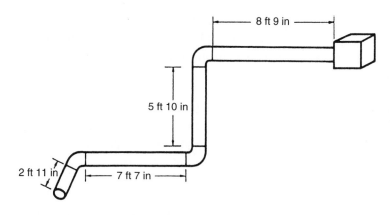

8 ft 9 in

5 ft 10 in

2 ft 11 in

7 ft 7 in

11. a. A bolt must pass through the layers shown. What is the thickness of the material to be bolted? Express the answer in centimeters.

_____

b. A 0.75-centimeter nut is to be screwed onto the bolt. The head of the bolt is 0.8 centimeter thick. What is the minimum length of the bolt to completely fill the nut?

_____

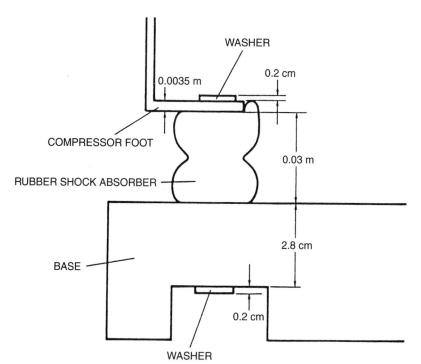

WASHER

0.2 cm

0.0035 m

COMPRESSOR FOOT

0.03 m

RUBBER SHOCK ABSORBER

BASE

2.8 cm

0.2 cm

WASHER

12. Pieces of copper tubing are used to install a hot water heating system. How many pieces, each 2 feet 4 inches long, can be cut from a 20-foot length of tubing?   _____

13. To repair a certain refrigerator, these lengths of wire are needed: 6 feet 4 inches, 2 feet 3 inches, 1 foot 11 inches, and 4 feet 9 inches. The lengths are cut from a 25-foot coil. Find in feet and inches the amount of coil left after the lengths are cut.   _____

14. Tape is used to seal insulation around 12 pieces of ducting. Each piece of duct needs two strips of tape. The first strip of tape measures 1.84 meters. The second strip measures 55 centimeters. The strips are cut from a 75-meter-long roll of tape. How much tape will be left after the 12 pieces of ducting are sealed?   _____

15. This rectangular duct is to be wrapped with insulation. How many inches of insulation are needed to completely wrap the duct?   _____

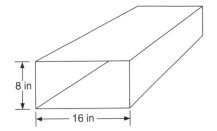

16. A band holds 1½-inch-thick insulation on the duct in problem 15. What is the minimum length of the band to go around the outside of the duct and have a 2-inch overlap?   _____

17. A magnetic strip fits around the grill of this air-conditioning unit. How long must the strip be to completely fit around the grill?   _____

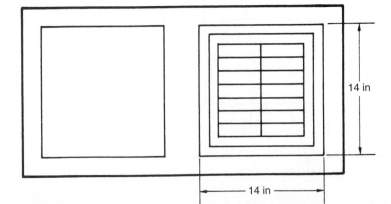

18. What length of insulation is needed to wrap this circular duct? Round the answer to the nearer hundredth inch.   _____

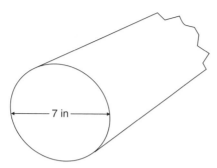

7 in

19. The thickness of the insulation in problem 18 is 1 inch. What length of tape would be needed to wrap around the outside of the insulation? (Include a 2-inch overlap for the tape.) Round the answer to the nearer hundredth inch.   _____

20. A refrigerator door is sealed with a magnetic gasket. The rectangular door is 36 inches wide and 39½ inches long. Find in feet and inches the total length of the gasket.   _____

21. This is an end piece for the support frame of an A-type evaporator coil. To force the airflow through the coil, weather stripping is needed. The stripping is measured along the center of the frame. How many inches of stripping are needed?   _____

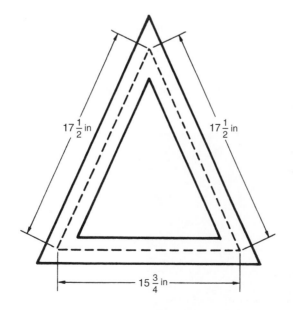

$17\frac{1}{2}$ in $17\frac{1}{2}$ in

$15\frac{3}{4}$ in

22. A compressor motor has a shaft diameter of ⅞ inch. The felt wiper that wraps tightly around the shaft must be replaced. What length of felt must be used? Round the answer to the nearer hundredth inch.   _____

23. Find to the nearer thousandth inch the circumference of the ring that fits around this piston for a compressor.   _____

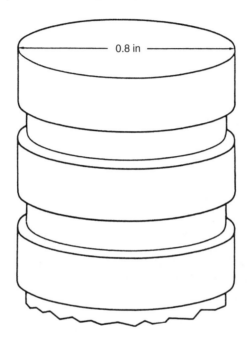

0.8 in

Note: Refer to the Section III of the Appendix on How to Read a Vernier or Micrometer for problems 24 and 25.

24. A piston ring from the piston in the previous problem needs to be replaced. Its thickness is measured using a vernier caliper. The vernier is shown. What is the thickness of the replacement ring?   _____

25. A shim broke and needs to be replaced. The thickness is measured using a micrometer. The micrometer is shown. What is the thickness of the needed shim?

_____

Computed Measure

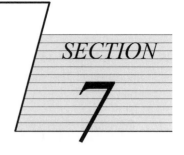

Unit 26 AREA MEASURE

BASIC PRINCIPLES OF AREA MEASURE

- Review the Introduction to Measurement in the front of the book.

- Review denominate numbers in Section I of the Appendix.

- Use formulas for areas located in Section III of the Appendix.

The amount of space on the surface of a figure is called the *area*. Area is also the number of square units equal in measure to the surface of a figure. Many figures have regular shapes to them. Formulas have been developed to find the area of these regular figures. The various formulas given below will be used throughout this unit.

To solve for area (*A*), determine the correct formula to use and then substitute values for the terms in the formula. Be certain that the units are the same. The units of area are square inches, square feet, square meters, or similar units.

> **Note:** When solving any of these problems, make a little table of each dimension and its measurement. Then write the correct formula, substitute into it as the next step, and then solve it. Be sure the units are all the same.

The areas for different figures can be found using the following formulas.

The area of a square is:

$$\text{Area} = \text{side} \times \text{side}$$
$$\text{or}$$
$$A = s \times s$$

$s \times s$ can also be written as s^2 (*s* squared).

So, $A = s \times s$

$$A = s^2$$

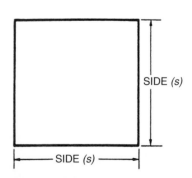

SIDE *(s)*

SIDE *(s)*

The area of a rectangle is:

Area = length × width
or
$$A = l \times w$$

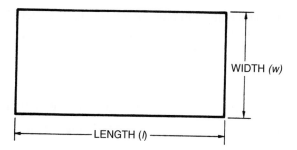

Note: We can find the length or the width of a rectangle if we know the area and the width or the length. We do this by using

$$l = \frac{A}{w} \quad \text{or} \quad w = \frac{A}{l}$$

The area of a circle is:

Area = π × radius × radius

$$A = \pi \times r \times r$$
$$A = \pi \times r^2$$

use π = 3.1416

$$A = 3.1416 \times r^2$$

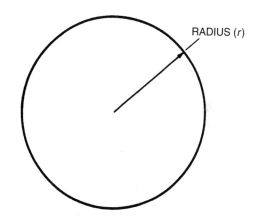

The area of a triangle can be found using:

$$\text{Area} = \frac{1}{2} \times \text{base} \times \text{height}$$
$$A = \frac{1}{2} \times b \times h$$

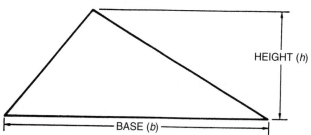

Note: The direction of *b* and *h* must form an angle of 90°. (They must be perpendicular to each other.)

We can also find either the base or the height of a triangle if the area is known and either the length or the height.

$$b = \frac{2 \times A}{h} \qquad h = \frac{2 \times A}{b}$$

Example 1: Find the area of a rectangle with a length of 6 inches (in) and a width of 2 inches.

$$l = 6 \text{ in} \qquad w = 2 \text{ in}$$

$A = l \times w$

$A = 6 \text{ in} \times 2 \text{ in}$

$A = 12 \text{ sq in}$

Example 2: A triangle has an area of 14 sq ft and a base of 7 feet. What is the height of the triangle?

$$A = 14 \text{ sq ft} \qquad b = 7 \text{ ft}$$

$$h = \frac{2 \times A}{b}$$

$$h = \frac{2 \times 14 \text{ sq ft}}{7 \text{ ft}}$$

$$h = \frac{28 \text{ sq ft}}{7 \text{ ft}} = 4 \text{ ft}$$

At times when finding the area of a figure, it must be broken into two figures.

Example 3: The wall shown has a round window in it. Find the area of the wall that must be insulated. Round the answer to the nearer hundredth.

Since the window is not insulated, to solve this problem, the area of the round window is subtracted from the area of the rectangular wall.

For the wall: $l = 18 \text{ ft} \qquad w = 8 \text{ ft}$

Estimating: $A = l \times w$

$= 20 \text{ ft} \times 8 \text{ ft} = 160 \text{ sq ft}$

Calculating: $A = l \times w$

$= 18 \text{ ft} \times 8 \text{ ft} = 144 \text{ sq ft}$

For the window: $D = 3 \text{ ft}$

$$r = \frac{D}{2} = \frac{3 \text{ ft}}{2} = 1.5 \text{ ft}$$

Estimating: $A = \pi \times r^2$

$= 3 \times (2 \text{ ft})^2 = 3 \times 4 \text{ sq ft} = 12 \text{ sq ft}$

Calculating: $A = \pi \times r^2$

= 3.1416 × (1.5 ft)² = 7.0686 sq ft or 7.07 sq ft

The wall area: Area of the rectangle − area of the circle

Estimating: A = 160 sq ft − 12 sq ft = 148 sq ft

Calculating: A = 144 sq ft − 7.07 sq ft = 136.93 sq ft

PRACTICAL PROBLEMS

1. The sides of a square duct are 8 inches. Find in square inches the area of the duct opening. _____

2. A square duct has sides of 24 centimeters. What is the area of the duct opening in square centimeters? _____

3. This grill at the end of an air duct has equally spaced square openings.

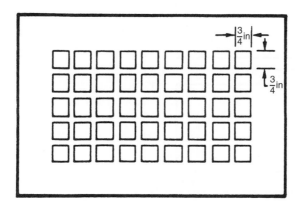

a. Find in square inches the area of each opening. _____

b. What is the total area of the openings through which air can flow? _____

4. Find in square meters the area of a wall that is 5 meters long and 3 meters high. _____

5.   The indicated wall must be painted after insulation is installed. Find in square feet the area of this wall so the proper amount of paint can be ordered.   _____

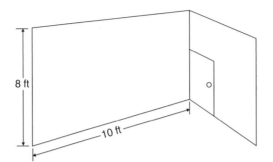

6.   What is the area of this window? Express the answer in square meters.   _____

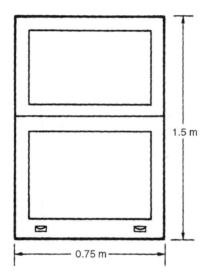

7.   A room in a house is heated with radiant matting. The room floorplan is shown. The portion that has the matting is the shaded part and is 18 inches away from each of the walls. What is the area of the heat matting in square feet?   _____

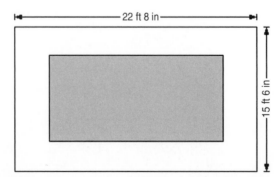

8. This house plan shows the dimensions of each room. Find in square meters the floor area of each room.

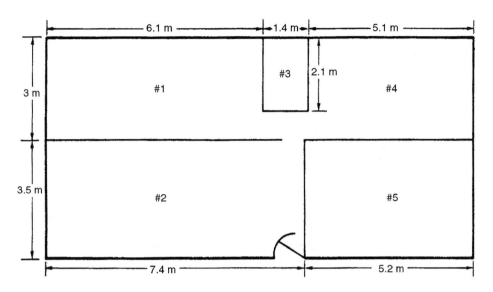

a. Room #1 _____

b. Room #2 _____

c. Room #3 _____

d. Room #4 _____

e. Room #5 _____

f. What is the total area of the house? _____

9. Styrofoam sheeting is used as insulation in the walls of refrigerators. Each wall is 0.55 meter by 1.6 meters. How many square meters of sheeting are needed for seven walls? _____

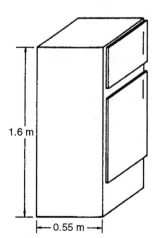

10. Find in square centimeters the area of the opening in this circular duct. Round the answer to the nearer thousandth.   _____

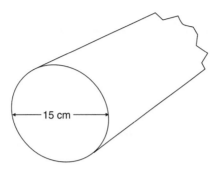

11. What is the area of the opening in a duct that has a diameter of 7 inches? Round the answer to the nearer thousandth square inch.   _____

12. The cylinder of a compressor has a diameter of 2.4 centimeters. What is the area of the opening in the cylinder? Round the answer to the nearer hundredth square centimeters.   _____

13. Find to the nearer hundredth square inch the area of the top of this piston for a compressor.   _____

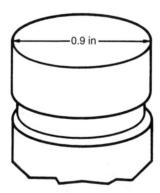

14. How many square centimeters of metal are needed to make this shield?   _____

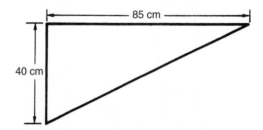

Note: The following figure of an A-frame building should be used for problems 15–17.

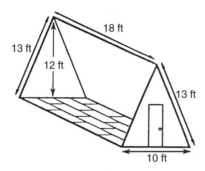

15. What is the area of the back triangular wall? _____

16. Find the area of the floor of this room. _____

17. What is the area of one of the slanting walls? _____

18. The front wall of the attic of one half of a duplex is in the shape of a triangle. What is the area of the front wall shown? _____

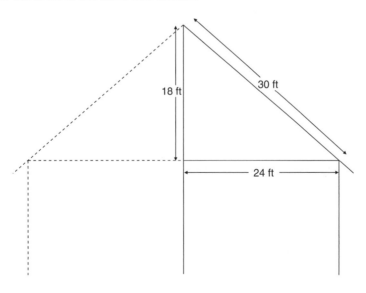

19. What is the area of this floor in square meters?   _____

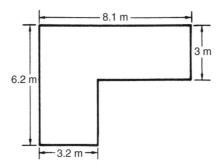

20. The end wall of a house is shaped like the figure shown. What is the area of the end wall of this house?   _____

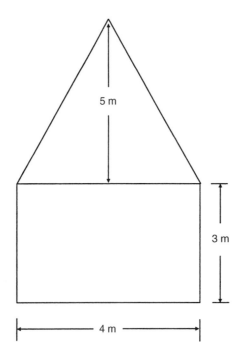

21. A rectangular duct must connect with a circular duct. The rectangular duct measures 6 inches by 7 inches. The circular duct has a diameter of 7 inches. Which duct has the smaller cross-sectional area?   _____

22. For proper airflow, a rectangular duct must have an area of 123.25 sq cm. The duct must be placed in the wall and must have a width of 8.5 centimeters. What must the length of the duct opening be?   _____

23. The area of a duct opening must be 84 sq in. The duct must be placed above a ceiling and must be 7 inches high. What length must the duct opening be?   _____

24. A windowpane made of glass for a hothouse added to the back of a house has 4.5 sq ft of glass. The pane has a length of 2.7 feet. What is the height of this pane? Round the answer to the nearer hundredth foot.   _____

25. An 8-inch by 12-inch rectangular duct splits into two branch ducts. The area of the two branches is equal to the area of the 8-inch by 12-inch duct. One of the branches is a square duct measuring 6 inches on each side. What is the area of the opening in the second branch?   _____

Unit 27 EQUIVALENT UNITS OF AREA MEASURE

BASIC PRINCIPLES OF EQUIVALENT UNITS OF AREA MEASURE

- Review denominate numbers in Section I of the Appendix.

- Study the tables of equivalent units of area measure in this unit and Section II of the Appendix.

- Review formulas for area in Section III of the Appendix.

Converting from one set of units of area measure to another is done in the same manner as conversion of linear measure. Fractions are formed to remove the current unit of measure and bring in a new unit. The only differences are the equivalent values.

Most everyone knows that there are 3 feet in 1 yard. The difficult question is, why aren't there 3 sq ft in 1 sq yd? It is easier to understand that idea if you look at a drawing of 1 sq yd. A sq yd is a square with each side having a length of 1 yard.

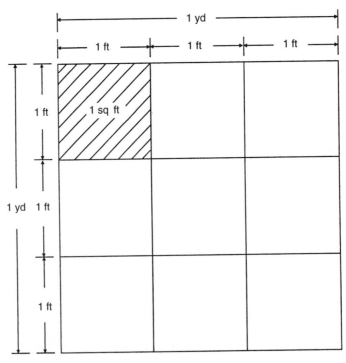

1 sq yd

A square with 1 yard on each side has an area of

$A = s^2$

$A = 1^2$

$A = 1$ sq yd

But as can be seen from the drawing, the area also contains 9 sq ft (not 3). This might be seen by finding the area of a square with each side having 3 feet.

$A = s^2$

$A = 3^2$

$A = 9$ sq ft

This is the reason that all new numbers must be used when changing from one set of units to another. These are not numbers that are normally known, so one should always refer to the tables.

The equivalent values are in the tables below.

ENGLISH AREA MEASURE		
1 square yard (sq yd)	=	9 square feet (sq ft)
1 square foot (sq ft)	=	144 square inches (sq in)

METRIC AREA MEASURE		
100 square millimeters (mm^2)	=	1 square centimeter (cm^2)
10,000 square centimeters (cm^2)	=	1 square meter (m^2)
1,000,000 square meters (m^2)	=	1 square kilometer (km^2)

When converting from English units to metric units or metric units to English units, use the following tables.

1 square meter (m^2)	=	10.763910 square feet (sq ft)
1 square meter (m^2)	=	1,550.000 square inches (sq in)
1 square centimeter (cm^2)	=	0.155000 square inch (sq in)
1 square millimeter (mm^2)	=	0.001550 square inch (sq in)

1 square foot (sq ft)	=	0.092903 square meter (m^2)
1 square inch (sq in)	=	0.000645 square meter (m^2)
1 square inch (sq in)	=	6.451600 square centimeters (cm^2)
1 square inch (sq in)	=	645.160 square millimeters (mm^2)

Note: When solving any of these problems, make a little table of each dimension and its measurement. Then write the correct formula, substitute into it as the next step, and then solve it. Be sure the units are all the same.

Example: A square measures 5 feet on each side. Find the area of the square in square inches. First, find the area in square feet.

$$s = 5 \text{ ft}$$
$$A = s^2$$
$$A = (5 \text{ ft})^2 = 25 \text{ sq ft}$$

Now convert 25 sq ft to sq in.

$$25 \text{ sq ft} \left(\frac{144 \text{ sq in}}{1 \text{ sq ft}} \right)$$

Estimating: $30 \times 100 = 3,000$ sq in
Calculating: $25 \times 144 = 3,600$ sq in

PRACTICAL PROBLEMS

1.   Convert 279 sq ft to sq yd.   _____

2.   The ceiling of a house has an area of 24 sq yd. What is the area in square feet?   _____

3.   To find the heat loss of a room, the area of this window must be determined. Find in square feet the area of this window.   _____

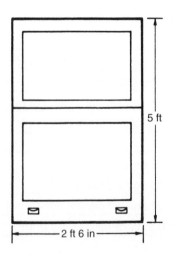

5 ft

2 ft 6 in

4. The freezer door of a side-by-side refrigerator needs to have its insulation
    replaced. What is the area of the insulation needed in square inches?   _____

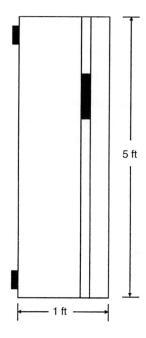

5 ft

1 ft

5. A roll of 3½-inch-thick insulation contains 30 feet of insulation. The insulation
    is 16 inches wide.

    a. Find the number of square inches of space the insulation will cover.   _____

    b. Find the number of square feet the insulation will cover.   _____

6. While repairing a baseboard hot water heating system, a technician drops a
    lighted blowtorch, badly damaging the rug. How many square yards of
    carpeting will be needed to replace the damaged rug? Round to the next
    higher full square yard.   _____

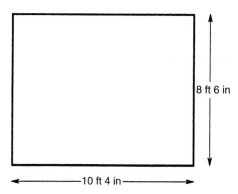

8 ft 6 in

10 ft 4 in

7.   The wall of a house has four windows. Each is a square window with a side of 82 centimeters. How many square meters of the wall are window?   _____

8.   A rectangular air duct measures 15 centimeters by 25 centimeters.

   a. What is the area of the opening in the duct in square centimeters?   _____

   b. What is the area of the opening in the duct in square meters?   _____

9.   A small number of buildings are built in the shape shown. They are known as flatiron buildings. One flatiron building has a roof with side lengths shown. What is the area of the roof of the building in square feet? Round to the nearer whole number. (**Hint:** Use the Pythagorean theorem.)   _____

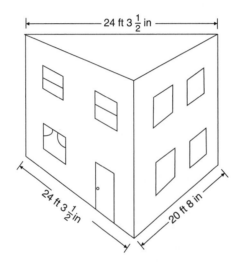

10. Find the area of a floor of a house built in the shape of an octagon (8 sides). (**Hint:** Divide the floor into 8 identical triangles.) Find the area in square meters.   _____

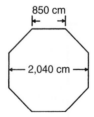

11. This filter is placed in a forced air system. What is the area through which the air can flow? Round the answer to the nearer hundredth square inch.   _____

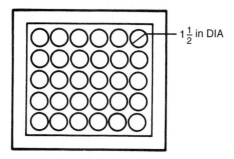

$1\frac{1}{2}$ in DIA

12. What is the area of the opening of a 7-inch round duct to the nearer tenth square foot?   _____

13. A circular air duct has a diameter of 21 centimeters. Find the cross-sectional area of the duct in square meters.   _____

14. This wall is to be insulated. Find the number of square feet of insulation that is needed. (**Note:** The window is not to be insulated, so its area is not included.)   _____

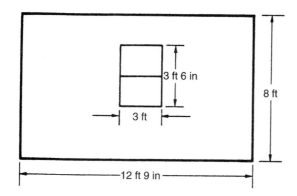

3 ft 6 in

8 ft

3 ft

12 ft 9 in

15. A heat load calculation is being done to determine how much heat is being lost through a wall. The area of the wall (not including the window) must be calculated in square feet. Find the area of the wall shown.

_____

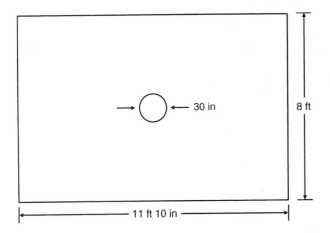

16. Find the area of the entranceway window shown. Express the area in square meters.

_____

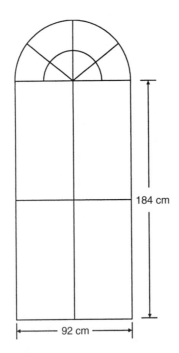

17. One of the 250 fins of a baseboard heater is shown. Each fin has a 2.5-centimeter hole in it.

    a. Find in square centimeters the area of the 250 fins.   _____

    b. Find in square meters the area of the 250 fins.   _____

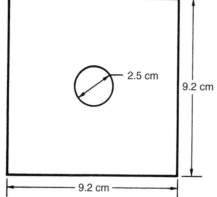

18. The ceiling in a room is sloped. What is the area of the wall in square feet?   _____

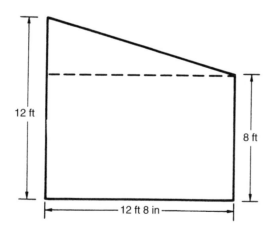

19. To reduce the heat load in a house during the summer, a reflective coating is applied to the two windows shown below. How many square inches of glass must be coated?   _____

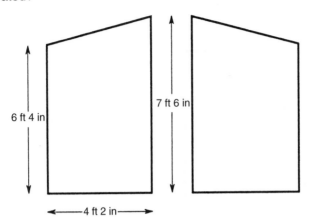

20. A different style of house is known as a saltbox.

   a. Find the surface area of an end wall of this variation of saltbox in square meters.   _____

   b. Find the area in square feet.   _____

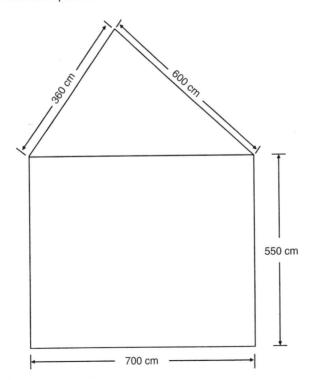

21. The opening in an air duct is 72 sq in. What is the area to the nearer hundredth square centimeter?

_____

22. A wall has an area of 75 sq ft. Find to the nearer hundredth square meter the area of the wall.

_____

23. The filter for a room air-conditioning unit has an area of 1,600 sq cm. How many square inches are there in the filter?

_____

24. The top of a piston for a compressor has an area of 3.4 sq in. Find the area to the nearer thousandth square centimeter.

_____

25. The installation instructions for an imported condensing unit for a domestic heat pump system state that it should sit on a slab at least 1.2 sq m in area. What is the minimum size of the slab in square feet? Round to the nearer tenth square foot.

_____

Unit 28 RECTANGULAR VOLUMES

BASIC PRINCIPLES OF RECTANGULAR VOLUMES

- Review the Introduction to Measurement in the front of the book.

- Review denominate numbers in Section I of the Appendix.

- Review the tables of equivalent units of length measure in Section II of the Appendix.

- Study the tables of equivalent units of volume measure in this unit and Section II of the Appendix.

- Use formulas for volumes located in this unit and in Section III of the Appendix.

Volume is the space enclosed by a three-dimensional figure. Volume is found by multiplying the length of a figure by its width and then multiplying that number by the figure's height. It can sometimes be found by multiplying the area of one surface by the depth of the figure from that surface. The units for volume are cubic units, such as cubic inches, cubic feet, or cubic meters.

A rectangular solid is a box-like figure. The volume (V) of a rectangular solid is:

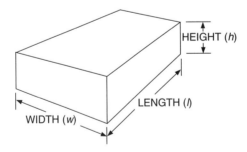

Volume = length × width × height

or

$$V = l \times w \times h$$

If the volume is known and one of the sides of the rectangular solid is not known, the formula can be rearranged to find that unknown value. If the length is not known, it can be found using:

$$l = \frac{V}{w \times h}$$

If the width is not known, it can be found using:

$$w = \frac{V}{l \times h}$$

If the height is not known, it can be found using:

$$h = \frac{V}{w \times l}$$

Again, care must be taken to make sure that all of the units are the same.

As seen in the figure below, a cubic yard is a cube with a length of 1 yard on each side. But each side is also equal to 3 feet. It can be seen, then, that 1 cu yd is equal to 27 cu ft.

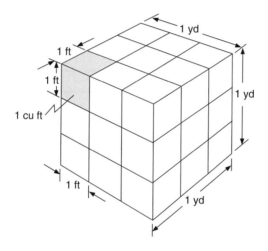

If units have to be changed, use equivalent units given in the tables below.

ENGLISH VOLUME MEASURE		
1 cubic yard (cu yd)	=	27 cubic feet (cu ft)
1 cubic foot (cu ft)	=	1,728 cubic inches (cu in)

METRIC VOLUME MEASURE		
1,000 cubic millimeters (mm^3)	=	1 cubic centimeter (cm^3)
1,000,000 cubic centimeters (cm^3)	=	1 cubic meter (m^3)

ENGLISH/METRIC VOLUME EQUIVALENCES		
1 cubic foot (cu ft)	=	28,317 cubic centimeters (cm^3)
1 cubic foot (cu ft)	=	0.028317 cubic meter (m^3)
1 cubic meter (m^3)	=	61,023 cubic inches (cu in)
1 cubic meter (m^3)	=	35.314667 cubic feet (cu ft)

Note: When solving any of these problems, make a little table of each dimension and its measurement. Then write the correct formula, substitute into it as the next step, and then solve it. Be sure the units are all the same.

Example: Find the volume of a rectangular room 12 feet long, 8 feet wide, and 8 feet high.

$l = 12$ ft

$w = 8$ ft

$h = 8$ ft

$V = l \times w \times h$

$V = 12$ ft $\times$ 8 ft $\times$ 8 ft

Estimating: $V = 10 \times 8 \times 8 = 640$ cu ft

Calculating: $V = 12 \times 8 \times 8 = 768$ cu ft

PRACTICAL PROBLEMS

1. Find in cubic feet the volume of this room.

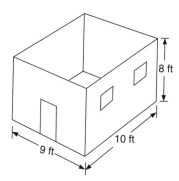

2. A mobile home measures 14 feet wide and 70 feet long. It has 7-foot-high ceilings. What is the approximate volume to be air-conditioned for this mobile home?

3. The inside of the back of a delivery truck has dimensions of 22 feet long, 8 feet wide, and 7 feet high. What is the maximum volume the truck is able to hold?

4. The condenser for a heat pump system sits on this slab of concrete. How many cubic centimeters of concrete are needed to make this slab?

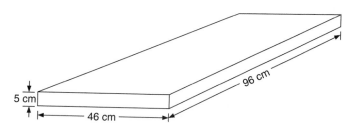

5.   Find in cubic meters the volume of this horizontal oil furnace.   _____

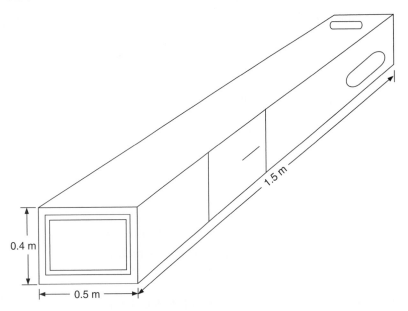

6.   The kitchen of a house measures 13 feet by 14 feet. There are 8-foot ceilings in the house. Find the volume in the room in cubic meters. Round to the nearer whole cubic meter.   _____

7.   Find the volume of the family room shown below. The room has a 10-foot-6-inch ceiling. Find the volume in cubic feet. Round to the nearer tenth cubic foot.   _____

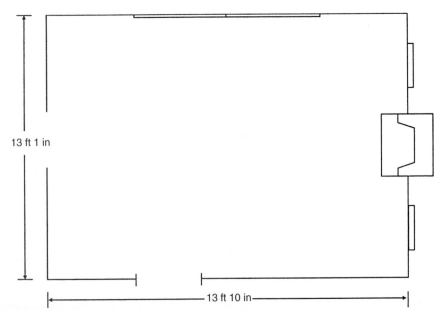

8. A homeowner is planning on adding cooling to the air conditioning for his house. What is the volume of space needed to be cooled?   _____

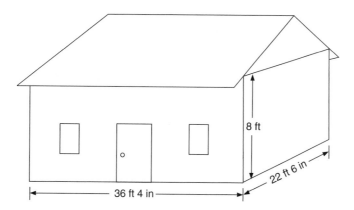

9. A warehouse has dimensions of 86 feet by 62 feet by 15 feet. How many cubic yards of supplies can this warehouse store? Round to the nearer cubic yard.   _____

10. The inside of a refrigerator measures 25 inches by 19 inches by 58 inches. What is the volume of the refrigerator in cubic feet? Round the answer to the nearer tenth cubic foot.   _____

11. The dimensions of this window air-conditioning unit are in inches.

   a. Find in cubic inches the volume of the unit.   _____

   b. What is the volume of the unit to the nearer hundredth cubic foot?   _____

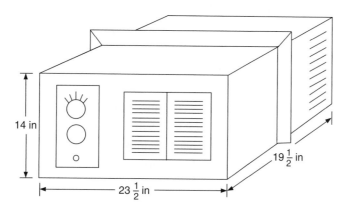

12. R-19 insulation is 6 inches thick. A roll of this insulation is 15 inches wide and has 32 feet of insulation in it. How many cubic feet of space will this insulation fill?   _____

13. The inside dimensions of a refrigerated tractor trailer are 91 inches across, 100⅞ inches high, and 44 feet ½ inch long. Find the volume in cubic feet that must be cooled by the refrigeration unit. Round the answer to the nearer tenth cubic foot.   _____

14. Find the volume of a 44-foot-8-inch by 30-foot by 8-foot house in cubic meters.   _____

15. An imported freezer lists its interior dimensions as 152 centimeters long, 65 centimeters wide, and 80 centimeters deep. What is the volume of the freezer in cubic feet? Round to the nearer hundredth cubic foot.   _____

16. Each story of a two-story house is 8 feet 6 inches high. The building is 38 feet long and 30 feet wide. Find the volume of the house.   _____

17. A room measures 12½ feet wide and 15½ feet long. The walls are 8 feet high. The volume of air in the room changes six times each hour. How many cubic feet of air enters the room each minute?   _____

18. The refrigerating compartment of a refrigerator has a volume of 11.55 cu ft. The inside of the compartment is 2.2 feet wide and 3 feet high. Find the inside length of the refrigerating compartment.   _____

19. If 1,350 cu ft of insulation are blown into an 1,800-sq ft attic, what is the thickness of the insulation in the attic?   _____

20. A room has a listed volume of 1,904 cu ft. The room is 8 feet high and 17 feet long. How wide is the room?   _____

21. Eight offices are on a common air-conditioning system. Each office is identical, measuring 12 feet by 10 feet 6 inches by 8 feet. What volume of air must the air handler move if it must change the entire volume of the offices five times each hour? Express the answer in cu yd per hour.   _____

22. The walls in this room are 8 feet high.

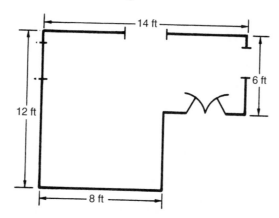

a. What is the volume of this room in cubic feet?

b. How many cubic meters of space are in this room? Round the answer to the nearer thousandth cubic meter.

_____

_____

23. Determine the volume of air in the ducts of a system if there are 72 feet of 8-inch by 12-inch duct and 45 feet of 6-inch by 10-inch duct in that system.

_____

24. In a particular area, a ⅓-ton cooling unit (4,000 Btu/hr [British thermal units per hour]) will comfortably cool an 18,600-cu ft house. A homeowner has a ⅓-ton unit for her 16,000-cu ft home. She would like to add a family room to the house. She would like it to be 20 feet by 15 feet by 8 feet. Can she build this addition without enlarging her cooling unit?

_____

25. A room in a warehouse measures 10 meters by 8.5 meters by 6 meters. What is the maximum number of rolls of insulation measuring 36 inches by 36 inches by 22¼ inches that can be stored in the room? Assume each roll is a rectangular block. Also assume that the space can be distorted, but the volume remains the same.

_____

Unit 29 *CYLINDRICAL VOLUMES*

BASIC PRINCIPLES OF CYLINDRICAL VOLUMES

- Review denominate numbers in Section I of the Appendix.

- Review the tables of equivalent units of length measure in Section II of the Appendix.

- Review the tables of equivalent units of volume measure in Section II of the Appendix.

Finding the volume of a cylindrical solid is similar to finding the volume of any solid, except a different formula is used. Cylindrical solids are solids with circular ends. The formula for finding the volume includes the area of a circle in it.

The volume (*V*) of a cylindrical solid is:

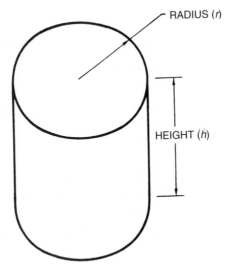

RADIUS (*r*)

HEIGHT (*h*)

$$V = \pi \times r \times r \times h$$
$$V = \pi \times r^2 \times h$$
Use $\pi = 3.1416$

This means that the volume of a cylindrical solid is:
$$V = 3.1416 \times r \times r \times h$$
$$V = 3.1416 \times r^2 \times h$$

Note: When solving any of these problems, make a little table of each dimension and its measurement. Then write the correct formula, substitute into it as the next step, and then solve it. Be sure the units are all the same.

Example: Find the volume of a cylinder 5 inches (in) in radius and 12 inches high.

$r = 5$ in $h = 12$ in

$V = \pi \times r^2 \times h$

$V = 3.1416 \times (5 \text{ in})^2 \times 12$ in

Estimating: $V = 3 \times 5^2 \times 10 = 750$ cu in

Calculating: $V = 3.1416 \times 5^2 \times 12 = 942.48$ cu in

PRACTICAL PROBLEMS

Note: Round all answers to three decimal places.

1. A cylindrical tank for refrigerant has an inside diameter of 14 inches and is 16 inches high. What is the volume of the tank in cubic inches?

    _____

2. The bore on a cylinder of a compressor is 2 inches. The stroke is 3 inches. What is the displacement of the compressor?

    _____

3. An addition to a house is shown below. What volume does this addition add to the house?

    _____

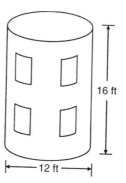

16 ft

12 ft

4. A fuel tank is a cylinder measuring 80 centimeters in diameter and 120 centimeters high. Find the volume of the fuel tank.

    _____

5. A compressor has a bore of 8 centimeters and a stroke of 10 centimeters. What is the displacement of the compressor?

    _____

6. Find in cubic centimeters the volume of this compressor.   _____

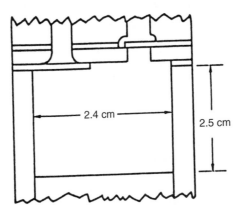

2.4 cm

2.5 cm

7. A cylinder containing propane has an inside diameter of 2.5 inches and is 10 inches long. How many cubic inches of propane can the container hold?   _____

8. A new city hall is built with a circular floor. The floor is 73 feet 6 inches in diameter. It is two stories high. Each story is 10 feet high. What is the volume of the building?   _____

9. Find the total volume of air in 44 feet of 6-inch round duct. Find the volume in cubic feet.   _____

10. What volume of refrigerant should be recovered for disposal from an old refrigerator? The length of ¼-inch tubing that has liquid in it is 15 feet long. Find the volume in cubic inches.   _____

11. An expansion tank for a domestic hot water system measures 7¾ inches in diameter and 20¼ inches long. What is the maximum volume the tank can hold?   _____

12. The hot water system in a house uses pipes that have an inside diameter (*ID*) of 1 inch. There are 25 feet of pipes in one part of the system. How many cubic feet of water can these pipes hold?   _____

13. Round ductwork measures 20 centimeters in diameter and is 14 meters long. What is the volume of this duct?   _____

14. An imported cylindrical tank has a radius of 35 centimeters and is 2.15 meters long. What is the volume of the tank?   _____

15. The combustion chamber of an oil furnace is a cylinder. The chamber is 18 inches long and has an inside diameter of 14 inches.

   a. What is the volume of the chamber in cubic inches?

   b. What is the volume of the chamber in cubic feet?

16. A load of 50 rolls of R-11 insulation must be picked up. Each roll is 25 inches in diameter and 15 inches tall. What is the smallest volume a closed-back truck could hold to be able to pick these rolls up in one load? (In reality, a larger volume is needed since the truck is rectangular and the rolls are cylindrical.) Round the answer to the next higher whole cubic foot.

17. Three cylindrical wells are dug as part of a ground source heat pump system. Each well is 6 inches in diameter and 95 feet deep. How many cubic feet of sand are needed to backfill the wells?

18. A tunnel in the shape of a half cylinder connects two buildings. The tunnel is 24 feet long and 10 feet 8 inches high at the highest point. What is the volume of air contained in the tunnel?

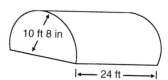

10 ft 8 in
|— 24 ft —|

19. This fuel tank is 60 inches long.

   a. Find in cubic inches the volume of the tank.

   b. One gallon of fuel will occupy 231 cu in of space. How many gallons of fuel will this tank hold?

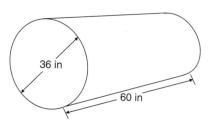

36 in
60 in

20. The inside diameter of the cylinder of the compressor for a heat pump is 5 centimeters. The cylinder is 3.65 centimeters long.

    a. Find in cubic centimeters the volume of the cylinder.   _____

    b. There is 0.061024 cu in of space in 1 cu cm of space. How many cubic inches of space are there in this compressor cylinder? Round the answer to the nearer tenth.   _____

21. A propane cylinder has a $5\frac{1}{2}$-inch radius and is 12 inches high. If 1 gallon is equivalent to 231 cu in, how many gallons of propane will the tank hold? (**Note:** Propane is usually measured by weight, but whatever the weight, what volume does this cylinder have?)   _____

22. A compressor has a 3-inch diameter piston with a variable stroke. The stroke can vary between 4 inches and $5\frac{1}{2}$ inches.

    a. What is the minimum volume the compressor will take in?   _____

    b. What is the maximum?   _____

23. Air is moving down an 8-inch round duct at a rate of 500 feet per minute. How much air flows out of the end of the duct each minute?   _____

24. An old cylindrical fuel oil tank is being removed from a worksite. It is 70 inches long and 40 inches in diameter. It is half full of *#2* fuel oil. Each cubic foot of fuel oil weighs 53.1 pounds. If the tank itself weighs 215 pounds, what is the total weight to be carried by the truck removing the tank? Round the answer to the nearer tenth pound.   _____

25. The inside dimensions of a rectangular oil furnace are 60 centimeters wide by 1.15 meters long by 45 centimeters high. The combustion chamber is cylindrical, is 46 centimeters long, and has an outside diameter of 43 centimeters. How many cubic centimeters of space are there between the inside walls of the burner and the outside walls of the combustion chamber?   _____

Formulas

8

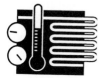

Unit 30 OHM'S LAW AND ELECTRICAL RELATIONSHIPS

BASIC PRINCIPLES OF OHM'S LAW AND ELECTRICAL RELATIONSHIPS

• Study and apply these principles of Ohm's Law to the problems in this unit.

Formulas are relationships that can be used over and over again by changing the values that are put into them. In formulas, letters represent numbers. The letters represent the same quantity—not the same value—all of the time. To solve the formula, you substitute numbers for those letters. Care must be taken to substitute the correct value for the correct letter. Care must also be taken to have the correct units with the values that are being substituted.

You have already been working with formulas. In the last several units, formulas were used to find areas and volumes. There are many different formulas that you will have to work with in the heating and air-conditioning field. In this unit, formulas dealing with electrical values will be studied.

Current is produced by electrons traveling from one point to another. It is the flow of electric charge. The unit measure for electrical current is *amperes*.

The flow of electrons is dependent upon the voltage of the system and the resistance. *Resistance* is the opposition to the flow of electric charge. It is measured in *ohms* (Ω). The *voltage* is the force applied to cause the electrons to flow through the resistance. It is measured in *volts*.

The relationship between the current (I), voltage (E), and the resistance (R) is known as *Ohm's Law*.

$$\text{current } (I) = \frac{\text{voltage } (E)}{\text{resistance } (R)}$$

The law states that the current (I) is directly proportional to the force (E) applied to produce the current and indirectly, or inversely, proportional to the resistance (R) to the flow of electrons. This means that the higher the voltage, the higher the current; and the smaller the resistance, the higher the current.

When solving problems using formulas, you want to have the quantity that you are trying to find all by itself on the left side of the equal sign. If the quantity you are looking for is on the left side of the equal sign by itself, then nothing has to be done.

However, if the formula is not set up correctly, then it must be manipulated or rearranged. Let us take some time and make sure we can manipulate formulas properly. The major rule to formula manipulation is to do the same thing to both sides of the equal sign. Often, it takes multiple steps to get the quantity you are trying to find to be by itself. Let us see how it is done.

Start with Ohm's Law

$$I = \frac{E}{R}$$

If we are trying to find the current (I), we do not have to do anything to the formula. If, however, we are trying to find the voltage (E), then we manipulate Ohm's Law as follows. We need to have E by itself. Since E is *divided by R, multiplying by R* will remove it from that side of the formula. BUT we must do the same thing to both sides of the formula!

$$I = \frac{E}{R}$$

$$I \times R = \frac{E}{R} \times R$$

$$I \times R = E$$

Now just exchange the two sides of the equal sign.

$$E = I \times R$$

Suppose we are asked to find R.

$$I = \frac{E}{R}$$

We need to get R in the numerator by itself. This is a two-step process. First multiply by R.

$$I \times R = \frac{E}{R} \times R$$

$$I \times R = E$$

Next we divide by I to remove it from the left side.

$$\frac{I \times R}{I} = \frac{E}{I}$$

$$R = \frac{E}{I}$$

The units for these electrical quantities are E in volts, I in amperes (or amps for short), and R in ohms.

$$E = I \times R$$

Volts = amperes (amps) × ohms

Another electrical term is power. *Power* is the rate of doing work. In electricity, it is the work done when one ampere of current is pushed through a circuit by a pressure of one volt. The unit of measure for power is the *watt* (W). The relationships between power, current, and voltage are:

power (P) = current (I) × voltage (E)

or

$$I = \frac{P}{E} \qquad E = \frac{P}{I}$$

The units used here are:

watts = amps × volts

Note: Always use the formula that has what you are looking for by itself on the left side of the equals sign.

An *electric circuit* is a conducting path that leads from the source, to energy devices or resistances, and back to the source. The current that flows through the circuit can travel in different paths.

When the electrical charge, or current, is provided with only one possible route to follow, the *circuit* is a *series circuit*. This means that the total current flows from the source, through each device, and back to the source.

The following electrical diagram is for a battery and three resistors wired in series.

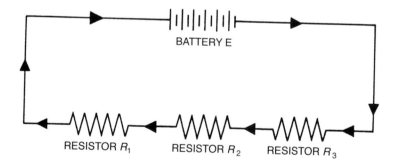

RESISTOR R_1 RESISTOR R_2 RESISTOR R_3

In a series circuit, the total resistance is equal to the sum of each device's resistance. For three resistors,

$$R_T = R_1 + R_2 + R_3$$

When the total current splits up and flows to the devices in separate paths, the circuit is a *parallel circuit.*

The circuit diagram for a battery and three resistors wired in parallel is shown below.

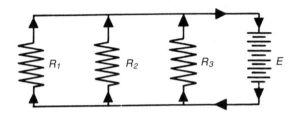

In a parallel circuit, the equivalent resistance for three resistors is:

$$R_T = \cfrac{1}{\cfrac{1}{R_1} + \cfrac{1}{R_2} + \cfrac{1}{R_3}}$$

In a series circuit, the total voltage is

$$E_T = E_1 + E_2 + E_3 + \ldots$$

but in a parallel circuit,

$$E = E_1 = E_2 = E_3 = \ldots$$

In a series circuit,

$$I = I_1 = I_2 = I_3 = \ldots$$

but in a parallel circuit,

$$I_T = I_1 + I_2 + I_3 = \ldots$$

Note: It is very important to realize that

$$R_T = \cfrac{1}{\cfrac{1}{R_1} + \cfrac{1}{R_2} + \cfrac{1}{R_3} + \ldots} \quad \text{is a lot different from } R_T = R_1 + R_2 = R_3 \ldots$$

To use the first formula, always solve it in the following way:

1.   Find each value for $\dfrac{1}{R_1}, \dfrac{1}{R_2}$, and so on. A calculator with a 1/x key simplifies this step.

2.   Add those values together. For many calculators, these two steps can be combined by putting in the number for R_1, pressing the 1/x key, pressing +, putting in the value for R_2, pressing the 1/x key, and so on, finally pressing the = key.

3.   Then divide that total into 1, or press the 1/x key again.

Some electrical circuits contain capacitors instead of resistors. A *capacitor* consists of two plates of electrical conducting material separated by an insulating material. *Capacitance* is the amount of electric charge a capacitor receives for each volt of applied potential. The unit of measure for capacitance is *microfarads*. For capacitors wired in parallel, the total capacitance is the sum of each capacitor's capacitance. The circuit diagram for three capacitors is shown below.

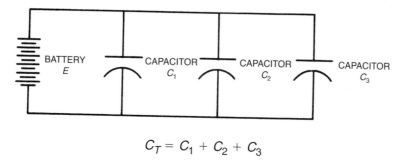

$$C_T = C_1 + C_2 + C_3$$

For capacitors wired in series, the equivalent capacitance for three capacitors is:

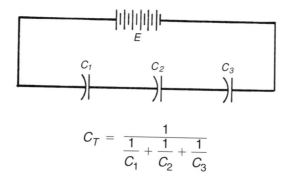

$$C_T = \cfrac{1}{\cfrac{1}{C_1} + \cfrac{1}{C_2} + \cfrac{1}{C_3}}$$

Note: This formula is solved in the same manner as

$$R_T = \cfrac{1}{\cfrac{1}{R_1} + \cfrac{1}{R_2} + \cfrac{1}{R_3}}$$

Example: A 4-ohm resistor is connected to a 12-volt battery. What current does it have flowing through it?

$R = 4$ ohms $E = 12$ volts

$$I = \frac{E}{R}$$

$$I = \frac{12 \text{ volts}}{4 \text{ ohms}} = 3 \text{ amps}$$

PRACTICAL PROBLEMS

Round the answer to the nearer hundredth when necessary.

Note: Use this diagram for problems 1–6.

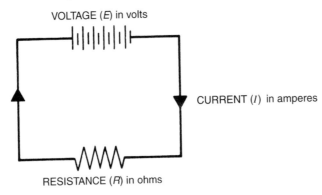

Complete this chart.

	VOLTAGE (E) in volts	CURRENT (I) in amperes	RESISTANCE (R) in ohms
1.	50	25	
2.		15	4
3.	25		10
4.	120	½	
5.	200		75
6.		5	72.5

7. A portable heater is listed for 115 volts. It uses 12 amperes of current. What is the resistance of the heater?   _____

8. An electric baseboard heater has 25 ohms of resistance. The heater uses 9.6 amperes of current. Find in volts the voltage required for the heater.   _____

9. The on-off switch light for a car air conditioner runs off the car's 12-volt battery. The light has a resistance of 16 ohms. What current passes through the bulb?   _____

10. The heater of the automatic defroster in a refrigerator is a resistance wire. The heater uses 110 volts and draws 4.8 amperes of current. Find in ohms the resistance.   _____

11. A window air-conditioning unit is rated at 230 volts and 13 amperes. What is the power used by the air conditioner?

_____

12. A 4,000-watt electric infrared quartz radiant heater is installed as a garage heater. It is wired into a 240-volt circuit. How much current will it draw when it is operating?

_____

13. A furnace for an electric heating system is rated at 121 amperes and 27,500 watts. What is the voltage of this system?

_____

14. A central air-conditioning unit is rated at 4,100 watts. The unit uses 230 volts. How many amperes of current does the electrical cable for this unit carry when current is flowing?

_____

15. A technician uses a 60-watt bulb in a portable light. This is plugged into a regular 120-volt household outlet. What current does the bulb have flowing through it when it is on?

_____

16. A portable drill has 400 milliamps of current when hooked to a 7.2-volt battery. What is the power output of the drill?

_____

17. A ½-horsepower (372.85-watt) compressor motor has 3.24 amperes of current flowing through it when running. What voltage is supplying the current to this motor?

_____

18. Three leads of a compressor are connected as shown. What is the total resistance between S and R when C is disconnected?

_____

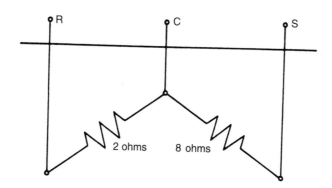

19. The left side, right side, and back panel heaters of a refrigerator are wired in parallel. Find the equivalent resistance of the circuit.

_____

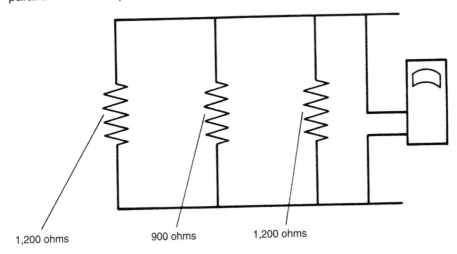

1,200 ohms   900 ohms   1,200 ohms

20. One room of a house has two mats for radiant heating. The mats are wired in parallel. One mat draws 5 amperes of current and the other draws 3 amperes. What is the effective current drawn by the mats in this room?

_____

21. The two electric baseboard heater units shown are placed in a room wired in series. What is the total resistance for the room?

_____

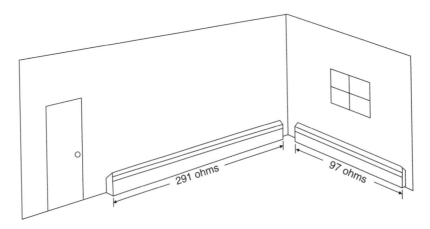

291 ohms   97 ohms

22. A technician must replace a faulty capacitor. The capacitor is replaced with two capacitors wired in parallel. What is the total capacitance?

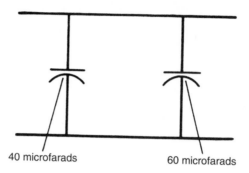

40 microfarads   60 microfarads

23. These relay capacitors are wired in series. Find in microfarads the total capacitance.

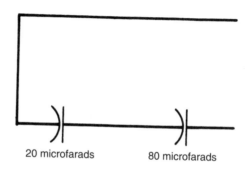

20 microfarads   80 microfarads

24. Wires act as resistance in circuits. This causes the voltage to decrease at the end of the wire. A building code states that a device should not be wired into a circuit if the voltage drop in the wires to that device is 5% of the rated voltage. A central air-conditioning system is to be wired into a 230-volt system.

a. What is the largest voltage drop in the wires that is allowed by the code?

b. If the starting current of the compressor is 0.089 ampere, find the resistance of the wire that would produce a 5% voltage drop in the wires to the compressor motor. Round to the nearer tenth ohm.

c. If #10 electrical cable has a resistance of $0.448 \frac{ohm}{foot}$, what is the maximum length in feet that the cable can be without producing the 5% voltage drop at the compressor? Round to the nearer whole foot.

25. Can two 25,000-Btu room air conditioners be wired into the same electrical circuit protected by a 25-ampere fuse? Each air conditioner is rated at 3,050 watts for a 230-volt circuit. (This rating is while the air conditioner is running steadily.)

Unit 31 GAS LAWS AND TEMPERATURE OF MIXTURES

BASIC PRINCIPLES OF GAS LAWS AND TEMPERATURE OF MIXTURES

- Study and apply these principles of gas laws and temperature of mixtures to the problems in this unit.

The majority of the problems in this unit deal with properties of gases. There are formulas relating the properties of gases to each other. The formulas are manipulated just as the formulas in the last unit were manipulated. Let us look at the properties of gases.

Fluid pressure is the force that a gas or a liquid exerts per unit area. It is expressed in pounds per square inch (psi). The metric unit of pressure is pascal (Pa).

$$1 \text{ psi} = 6,895 \text{ Pa}$$

The pressure of a gas is usually measured by using a pressure gauge. A pressure gauge usually reads zero under normal atmospheric conditions. A gauge pressure reading is given with units of pounds per square inch gauge (psig).

The air also exerts a pressure. This pressure is called *atmospheric pressure*. Atmospheric pressure at sea level is 14.7 psi.

Absolute pressure uses both the gauge pressure reading and the atmospheric pressure. It is measured in pounds per square inch absolute (psia).

$$\text{absolute pressure} = \text{gauge pressure} + 14.7 \text{ psi}$$

The pressure of a gas is dependent upon temperature and the volume of the gas. A change in one of these three properties of the gas will result in a change in one or both of the others. The relationship between these quantities is known as the *general law of perfect gas.*

$$\frac{P_1 \times V_1}{T_1} = \frac{P_2 \times V_2}{T_2}$$

or $\quad P_1 \times V_1 \times T_2 = P_2 \times V_2 \times T_1$

where $\quad P_1$ = original absolute pressure

V_1 = original volume

T_1 = original absolute temperature

P_2 = new absolute pressure

V_2 = new volume

T_2 = new absolute temperature

If any of the properties of the gas remain constant throughout the process, the 1 value will equal the 2 value in the formula, and they can be canceled. This simplifies the formula, making it easier to solve.

If the temperature of a gas remains constant, then T_1 will equal T_2, and the formula will simplify to

$$P_1 \times V_1 = P_2 \times V_2$$

This relationship is known as *Boyle's Law.* We say that a change in the pressure of the gas is indirectly, or inversely, proportional to the volume of the gas. This means that if the pressure of the gas doubles, the new volume occupied by the gas is one-half the original volume. If the new volume is four times the original volume, the new pressure is one-fourth the original pressure.

If the volume of the gas remains constant, V_1 equals V_2, and the general law simplifies to

$$\frac{P_1}{T_1} = \frac{P_2}{T_2}$$

or $\qquad P_1 \times T_2 = P_2 \times T_1$

This law is known as *Charles' Law.* We say that a change in the pressure of the gas is directly proportional to the temperature of the gas. This means that if the pressure of the gas doubles and the volume of the gas stays the same, the temperature of the gas doubles. If the new temperature is four times the original temperature, the new pressure will be four times the original pressure.

If the pressure of the gas remains constant, P_1 equals P_2, and the general law simplifies to

$$\frac{V_1}{T_1} = \frac{V_2}{T_2}$$

or $\qquad V_1 \times T_2 = V_2 \times T_1$

This law is known as *Guy-Lussac's Law.* We say that a change in the volume of the gas is directly proportional to the temperature of the gas. This means that if the absolute temperature of the gas doubles and the pressure of the gas remains the same, the volume of the gas will also double.

An important point to remember is that each of these last three laws is just a simplification of the general law. You can learn just one law, the general law, and solve all of the problems associated with gases. The last three laws, however, can make it a little easier to solve some of the problems. It should also be noted that the last two laws are, in reality, proportions that you have already learned to solve.

In each of these laws, the pressure and temperature must be absolute! *P* and *T* must both be in absolute units, in other words, pounds per square inch absolute (psia) for pressure and Kelvins or degrees Rankine (K or °R) for temperature, not pounds per square inch gauge (psig) pressure or degrees Fahrenheit (°F) or degrees Celsius (°C) for temperature. When working gas law problems, pressure and temperature must be converted to absolute before working the problems. They may have to be converted back after the value has been found, depending upon what is asked in the problem.

- Study this table of formulas for finding original and new pressures, temperatures, and volumes.

GAS LAW	ORIGINAL VALUE	NEW VALUE
General Law of Perfect Gas $$\frac{P_1 \times V_1}{T_1} = \frac{P_2 \times V_2}{T_2}$$ or $$P_1 \times V_1 \times T_2 = P_2 \times V_2 \times T_1$$	$$P_1 = \frac{P_2 \times V_2 \times T_1}{V_1 \times T_2}$$ $$V_1 = \frac{P_2 \times V_2 \times T_1}{P_1 \times T_2}$$ $$T_1 = \frac{P_1 \times V_1 \times T_2}{P_2 \times V_2}$$	$$P_2 = \frac{P_1 \times V_1 \times T_2}{V_2 \times T_1}$$ $$V_2 = \frac{P_1 \times V_1 \times T_2}{P_2 \times T_1}$$ $$T_2 = \frac{P_2 \times V_2 \times T_1}{P_1 \times V_1}$$
Boyle's Law $$P_1 \times V_1 = P_2 \times V_2$$	$$P_1 = \frac{P_2 \times V_2}{V_1}$$ $$V_1 = \frac{P_2 \times V_2}{P_1}$$	$$P_2 = \frac{P_1 \times V_1}{V_2}$$ $$V_2 = \frac{P_1 \times V_1}{P_2}$$
Charles' Law $$\frac{P_1}{T_1} = \frac{P_2}{T_2}$$ or $$P_1 \times T_2 = P_2 \times T_1$$	$$P_1 = \frac{P_2 \times T_1}{T_2}$$ $$T_1 = \frac{P_1 \times T_2}{P_2}$$	$$P_2 = \frac{P_1 \times T_2}{T_1}$$ $$T_2 = \frac{P_2 \times T_1}{P_1}$$
Guy-Lussac's Law $$\frac{V_1}{T_1} = \frac{V_2}{T_2}$$ or $$V_1 \times T_2 = V_2 \times T_1$$	$$V_1 = \frac{V_2 \times T_1}{T_2}$$ $$T_1 = \frac{V_1 \times T_2}{V_2}$$	$$V_2 = \frac{V_1 \times T_2}{T_1}$$ $$T_2 = \frac{V_2 \times T_1}{V_1}$$

Remember: Be sure that P and T both are in absolute units. In other words, use psia and K or °R, not psig and °F or °C. Refer to Unit 21 or Section IV of the Appendix to convert temperatures.

Always use the formula that has what you are looking for by itself on the left side of the equals sign.

The remainder of the problems deal with the temperature of mixtures. The temperature of mixtures is related to the conservation of energy (or heat). When two or more amounts of material at different temperatures mix, the energy (or heat) is spread between all of the material so that everything comes to a uniform temperature. The formulas for these problems are included with the individual problems.

PRACTICAL PROBLEMS

Round the answer to the nearer hundredth when necessary.

Complete this chart.

	P_1	V_1	T_1	P_2	V_2	T_2
1.	30 psia	20 cu in	constant	50 psia	___ cu in	constant
2.	25 psia	14 cu in	constant	___ psia	21 cu in	constant
3.	15 kPa	10 cu ft	constant	___ kPa	20 cu ft	constant
4.	36 psig	8 cu in	constant	48 psig	___ cu in	constant
5.	20 psia	constant	27°C	___ psia	constant	77°C
6.	10 psia	constant	17°C	15 psia	constant	___°C
7.	12 psia	49 cm³	77°C	___ psia	49 cm³	59°C
8.	constant	10 cm³	40°C	constant	___ cm³	80°C
9.	constant	20 cu in	80°F	constant	40 cu in	___°F
10.	24 psia	2 cu in	27°C	___ psia	1.3 cu in	52°C
11.	9 psia	5 cu in	22°C	12 psia	4.5 cu in	___°C
12.	16 psig	6 cu in	70°F	30 psig	___ cu in	60°F

13. An air compressor begins its cycle with 0.8 cu in of air at atmospheric pressure (14.7 psi or 0 psig) in its cylinder. The air leaving the cylinder has an absolute pressure of 42 psia. The temperature remains the same. What is the new volume of the air leaving the compressor?

_____

14. A compressor takes 1.2 cu in of gas at a pressure of 14.7 psia and compresses it into a volume of 0.36 cu in. If the temperature of the gas remains the same, find the pressure of the gas after it is compressed.

_____

15. An oxygen cylinder for an oxyacetylene setup registers a pressure of 1,724 kPa in the afternoon when the technician is finished using it. The temperature of the cylinder in the afternoon is 30°C. In the morning the temperature is 20°C and the cylinder registers a pressure of 1,668 kPa. Has the cylinder developed a leak?

_____

16. Air passing through a heater has its temperature raised from 68°F to 98°F. A cubic foot of air undergoes this heatup. What is the new volume of this air? (**Note:** Since the volume is larger, the air is less dense. This decrease in density is what causes warm air to rise.)

17. A large electric generator is cooled by a gas that then passes through a heat exchanger and is cooled itself. One cubic meter of gas enters the heat exchanger with a temperature of 77°C. When it leaves the heat exchanger, it occupies 0.95 cu m. What is the temperature of the gas as it leaves the heat exchanger?

18. In the morning, a tank containing refrigerant R-134a as a gas is at a pressure of 60 psia. The temperature is 60°F. In the afternoon, the temperature reaches 95°F. What is the new pressure inside the tank?

19. Find the exhaust pressure for the compressor.

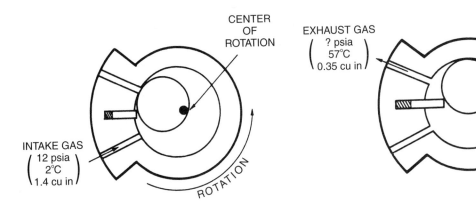

20. An expansion valve in an improperly operating refrigerator system is shown below. The system is misoperating such that there is gas on both sides of the expansion valve, not liquid as would be found on a properly operating system. What would a pressure gauge read on the low pressure side of the expansion valve?

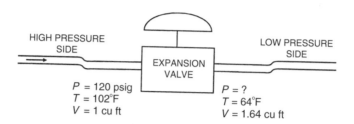

HIGH PRESSURE
SIDE

EXPANSION
VALVE

LOW PRESSURE
SIDE

$P = 120$ psig
$T = 102°F$
$V = 1$ cu ft

$P = ?$
$T = 64°F$
$V = 1.64$ cu ft

21. A nitrogen cylinder develops a leak. The gas inside the cylinder is at a pressure of 250 psia and room temperature 72°F. When it leaks, 1 cu in of nitrogen inside the tank will expand to 15.35 cu in in the atmosphere. What is the temperature of the nitrogen as it leaks out of the cylinder?

Note: Use this diagram for problems 22 and 23.

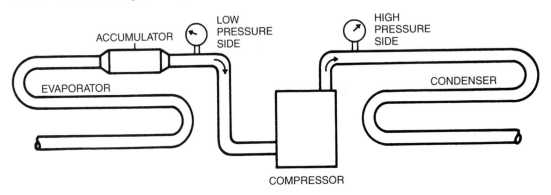

22.   The values for the evaporator and condenser side of the air conditioner compressor system are given. Find in Kelvins the temperature of the condenser side.   _____

EVAPORATOR SIDE   CONDENSER SIDE

P = 12.5 psia P = 174 psia

V = 1.2 cu in V = 0.11 cu in

T = 245K T = ?

23.   The values for the low pressure and high pressure side of this air conditioner compressor system are given. Find the missing value.   _____

LOW PRESSURE SIDE   HIGH PRESSURE SIDE

P = 0.3 psig P = 137.3 psig

V = 0.95 cu in V = ?

T = 3°C T = 47°C

24. A hydronic heating system has hot water leaving the boiler and then splitting to flow through two radiators. The radiators can give up different amounts of heat. They can also have different flow rates through the radiators. The temperature of the water returning to the source of heat is determined by the mixture of the waters leaving the radiators and can be found using the formula:

> Flow 1 × Temperature 1 + Flow 2 × Temperature 2
> = Total flow × Temperature return

What is the return temperature for the system shown?

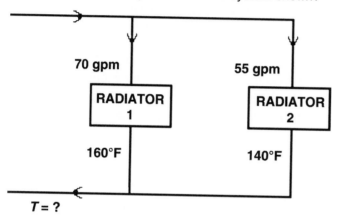

25. Three radiators in a hydronic system are very similar to two radiators in a system. You add a third product of Flow × Temperature to the left side of the equation before solving. What is the return temperature for the system shown?

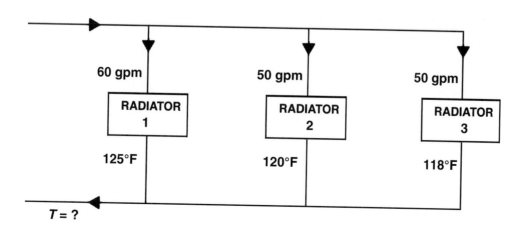

Unit 32   HEAT LOAD CALCULATIONS

BASIC PRINCIPLES OF HEAT LOAD CALCULATIONS

- Review and apply the principles of formula manipulation to these problems.

One activity that a heating and cooling technician may have to perform is determining the size of a heating or cooling system. Too small a system will make the customer unhappy because the house is too hot or too cool, and too large a system will cost more than a proper system would cost. It is important to size the unit correctly. To get the correct size unit, the proper heat load must be determined.

The heat load is the amount of heat that would be lost or gained each hour with the extreme temperature difference for which the system is designed. Heat loads are not calculated for interior structures, since no heat is lost through these structures. Heat loads are determined by multiplying the heat transfer multiplier (the amount of heat transferred through one square foot of the structure) for a type of structure by the area of that structure. The formula is written as

$$\text{heat load} = \text{heat transfer multiplier} \times \text{area}$$

A wall is often made up of a number of different components, such as a window, plain wall, and a door, to name a few. When calculating the heat load for the wall, you need to calculate the heat load for each component separately and then add these results together. The area of each component must be found and the proper heat transfer multiplier must be used for that component. The last step is to add all of the heat load values together.

The cooling load for a building is found using the same formula that was used to find the heat load. For a cooling load, the design temperature difference is usually 25°F instead of the 70°F or 75°F used for the heat loads. There is an additional cooling load created when the sun shines into a room. If the room does not get direct sunlight, this additional load is not put on the cooling load. A window facing south would allow a maximum of 75 Btu per hour to enter the room for each square foot of window area. This must be added to the calculated cooling load due to the temperature difference.

Use the table on page 200 or the one found in Section V of the Appendix to get the correct heat transfer multipliers.

Note: Always use the formula that has what you are looking for by itself on the left side of the equals sign.

Example: Find the heat load for the following side of a house for a 70°F design temperature difference. The wall is a wood frame with siding and 3½ inches of insulation. The window is double pane.

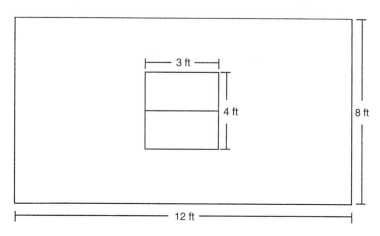

The window:

$$A = l \times w$$

$$A = 3 \text{ ft} \times 4 \text{ ft} = 12 \text{ sq ft}$$

Heat load $=$ Heat transfer multiplier $\times$ area

$$\text{Heat load} = 70 \, \frac{\text{Btu}}{\text{h} \cdot \text{sq ft}} \times 12 \text{ sq ft} = 840 \, \frac{\text{Btu}}{\text{h}}$$

The wall:

$$A = l \times w - \text{area of window}$$

$$A = 12 \text{ ft} \times 8 \text{ ft} - 12 \text{ sq ft} = 96 \text{ sq ft} - 12 \text{ sq ft}$$

$$A = 84 \text{ sq ft}$$

Heat load $=$ Heat transfer multiplier $\times$ area

$$\text{Heat load} = 5 \, \frac{\text{Btu}}{\text{h} \cdot \text{sq ft}} \times 84 \text{ sq ft} = 420 \, \frac{\text{Btu}}{\text{h}}$$

The entire side of the house:

$$\text{Heat load} = 840 \, \frac{\text{Btu}}{\text{h}} + 420 \, \frac{\text{Btu}}{\text{h}} = 1{,}260 \, \frac{\text{Btu}}{\text{h}}$$

HEAT TRANSFER MULTIPLIERS

Note: The heat transfer multiplier is found by multiplying the U factor (the amount of heat transferred through 1 sq ft of structure for each degree temperature difference between the inside and outside surfaces) by the design temperature difference. The units for the heat transfer multiplier are British thermal units per hour per square foot, so all areas must be in units of square feet.

TYPE OF STRUCTURE	DESIGN TEMPERATURE DIFFERENCE		
	25°F	70°F	75°F
Walls — wood frame with sheathing and siding or other veneer			
3½ inches insulation (R-11)	3.5	5	5
3½ inches insulation + 1 inch polystyrene sheathing (R-16)	3.1	3.5	3.8
6 inches insulation (R-19)	2.6	2.8	2.9
Brick with 3½ inches insulation	3.4	4.5	4.9
Brick with 3½ inches insulation + 1 inch polystyrene sheathing	3.0	3.4	3.7
Metal siding with 1½ inches polystyrene	3.3	4.3	4.7
Metal siding with 4 inches polystyrene	1.3	3.3	3.6
Ceiling — under vented roof			
3½ inches insulation (R-11)	2.5	6	6
6 inches insulation (R-19)	1.5	4	4
9½ inches insulation (R-30)	1.0	2.2	2.4
Floor			
Slab on ground	No heat loss	No heat loss	No heat loss
No insulation over crawl space/basement	5	16	17
6 inches insulation over crawl space/basement	1	3.2	3.4
Windows			
Single pane	35	105	110
Double pane	25	70	75
Single pane + storm window	25	60	65
Double pane (fixed)	25	60	65
Double pane + storm window	12	35	38
Doors			
Insulated core, weather-stripped	5.3	81	86
Sliding glass door, double glass	25	90	95

PRACTICAL PROBLEMS

Express the answer to the nearer British thermal unit per hour.

1. A warehouse measures 40 feet by 50 feet and has 20-foot-high walls. The warehouse was built on a concrete slab and has 6 inches of insulation in the wood frame walls and 9½ inches in the ceiling. There are no windows in the building, and the door is made just like the walls. What is the heat load for this warehouse in an area where there is a 75°F design temperature difference?

_____

2. A house is built in a location with a design temperature difference of 75°F. The room shown is a corner room on the second story, so it loses heat only through the two outside walls and the ceiling. The walls are wood frame with sheathing and siding and 3½ inches of insulation (R-11). The windows are single pane. The ceiling has 6 inches of insulation (R-19) under a vented attic. Find the heat load for this room.

_____

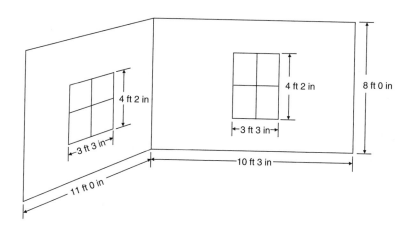

3. A one-story house is built over a vented crawl space. The attic is also vented. The heating system is designed for a temperature difference of 75°F. The exterior wall of a middle bedroom is shown. The room is 10 feet 3 inches wide. The window has double pane glass. The wall is wood frame with sheathing and siding. It has 1 inch of polystyrene insulation over 3½ inches of insulation (total R-16). The ceiling has 9½ inches of insulation (R-30), and the floor has 6 inches of insulation (R-19). Find the heat load for this room.

_____

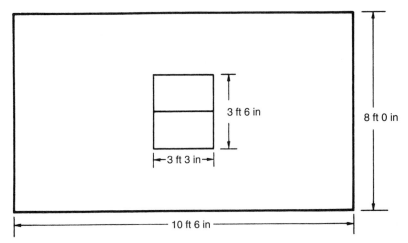

4. A dining room of a two-story house is located in the corner of the house. The house is designed for a temperature difference of 70°F. The dining room has a sliding glass door with double glass. The walls are wood frame with 3½ inches of insulation (R-11) and sheathing and siding. The floor has 6 inches of insulation (R-19) and is over an unheated basement. This basement gives the floor a design temperature difference of 25°F.

a. What is the heat load for this room?

_____

b. What percent of the heat is lost through the sliding glass door? Express the answer to the nearer tenth percent.

_____

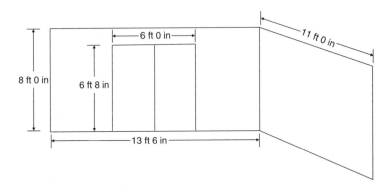

5. A corner bedroom is situated on the second story of a two-story house. The house is designed for a temperature difference of 70°F. Above the bedroom is a vented attic. There are 9½ inches of insulation (R-30) on the ceiling. The walls are frame with sheathing and siding and have 1 inch of polystyrene insulation over 3½ inches of insulation (total R-16) in them. The windows are single pane. Find the heat load for this room.

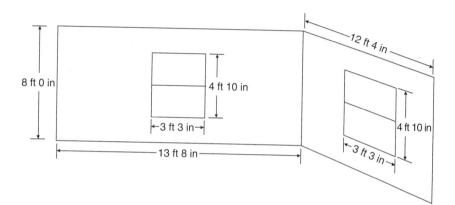

6. Find the heat load for the room in problem 5 if storm windows were installed.

7. A house is designed for a temperature difference of 75°F. The living room has an insulated core door that is weather-stripped. The windows are double pane with the center section fixed. This living room is one corner of the first floor of a two-story house and is over a vented crawl space. The floor has 6 inches of insulation (R-19) in it. The side wall is frame and has 3½ inches of insulation plus one inch of polystyrene insulation (total R-16) on it. The front wall has brick veneer over the same type of insulation (R-16). Find the heat load for this room.

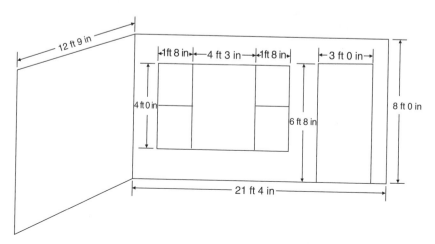

8. Find the heat load for an upstairs 14-foot by 12-foot bedroom that is not on the corner of the house. The outside wall, the 14-foot wall, has $3\frac{1}{2}$ inches of insulation. The walls are 8 feet high, and there are two single pane windows in that outside wall that are 4 feet 5 inches high by 3 feet wide. The ceiling has insulation 6 inches thick. The house has a design temperature difference of 70°F.

    _____

9. The room described in problem 8 is remodeled. One-inch polystyrene sheathing has been added to the outside wall. The ceiling now has $9\frac{1}{2}$ inches of insulation. The windows are double pane. What is the new heat load for this room?

    _____

10. The owner of the room in problem 9 bought a storm window to add to the current window. How much will that reduce the head load?

    _____

11. The walk-in freezer shown has walls, floor, and ceiling insulated with 4 inches of polystyrene. A restaurant is deciding whether to put it inside the building or put it as a self-standing unit outside. How much more cooling will be required for the unit in summer outside compared to inside? Assume an outside temperature difference of 75°F and an inside temperature difference of 25°F. Assume no heat enters through the floor.

    _____

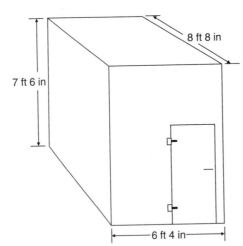

8 ft 8 in

7 ft 6 in

6 ft 4 in

12. A corner office measures 9 feet by 10 feet 6 inches with an 8-foot ceiling. The two outside walls are brick with $3\frac{1}{2}$ inches of insulation. There are no windows in the walls. The ceiling has $9\frac{1}{2}$ inches of insulation beneath a vented roof. The floor is a slab floor sitting on the ground. What is the heat load for this office for a design temperature difference of 75°F?

    _____

13. It is desired to put a 5-foot-high by 3-foot 6-inch double pane (fixed) window in each of the two outside walls of the office described in problem 12. What will be the new heat load?

    _____

14. How much heat gets into the refrigerator/freezer each hour? Assume room temperature is 72°F. Also assume that the doors are not opened. Assume the refrigerator is kept at 47°F and the freezer is kept at 2°F. All of the walls including the roof and floor are composed of $1\frac{1}{2}$ inches of polystyrene.

    _____

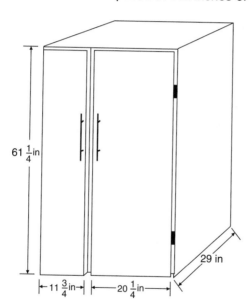

15. A one-story house built over a vented crawl space is designed for a 70°F temperature difference. The attic area is vented. The walls all have 3½ inches of insulation (R-11) with sheathing and siding. The ceilings have 9½ inches of insulation (R-30), and the floor has 6 inches (R-19). All windows are double pane, and each one is 4 feet 2 inches high and 3 feet 3 inches wide. The sliding glass door is 6 feet 8 inches high and 6 feet wide. Both doors are 6 feet 8 inches high and 3 feet wide, have insulated cores, and are weather-stripped. The ceilings are 8 feet high. Find the heat load for the house.   _____

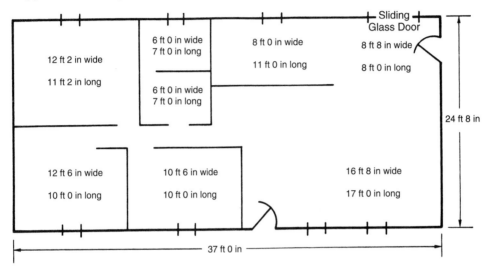

16. The cooling load for a house is determined in part by using the same formula used to find the heat load. The temperature difference used in designing a cooling system is 25°F. The corner room shown does not face the sun so no additional cooling load is required. The windows are single pane, and the walls are brick veneer with 3½ inches of insulation (R-11). The ceiling has 3½ inches of insulation (R-11), and the floor has no insulation. The room is built over a vented crawl space with a vented attic above it. Find the cooling load for this room.   _____

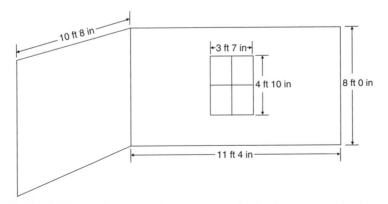

17. If the room in problem 16 faced the sun, an additional cooling load would be created. This extra cooling load would be due to the sun shining through the window and heating the room. A window facing south would allow a maximum of 75 Btu per hour to enter the room for each square foot of window area. If the room has its window facing south, what is the maximum cooling load for the room?

18. Determine the cooling load for the house shown. Each window is 4 feet high and 3 feet wide made of double pane glass. Each door is 7 feet high and 4 feet 6 inches wide with an insulated core, weather-stripped. There are no windows in the doors. The sun strikes the back of the house. The brick wall exterior of the house has $3\frac{1}{2}$ inches of insulation. The walls are 8 feet high. The roof has $9\frac{1}{2}$ inches of insulation and a vent. The house is built on a slab.

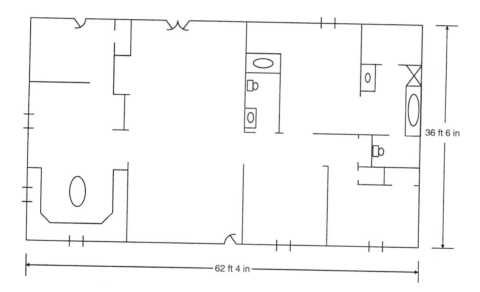

19. A restaurant wants to change the design of a dining room. One of its solid walls is being replaced with a fixed double pane glass wall. The design temperature difference is 70°F. The wall measures 9 feet high and 20 feet 6 inches long. It was a brick wall with $3\frac{1}{2}$ inches of insulation. What will be the additional heat load for the ventilation system during the winter due to this wall?

20. For the dining room described in problem 19, what will be the additional cooling load during the summer? The design temperature difference is 25°F. The new glass wall will face the sun.

21.   A two-story house has its floor plan shown. There is an unheated basement. The top-story ceiling and first-story floor have R-19 insulation. The walls have 3½-inch insulation and are 8 feet high. All of the windows are single pane with storm windows. Each window measures 3½ feet wide and 4½ feet high. The doors are insulated core, weather-stripped ones. The sliding glass doors are double pane. All regular doors measure 6 feet 4 inches high and 3 feet wide. The sliding glass door measures 6 feet 4 inches high and 6 feet wide. For a 70°F design temperature difference, what is the heat load for the house? _____

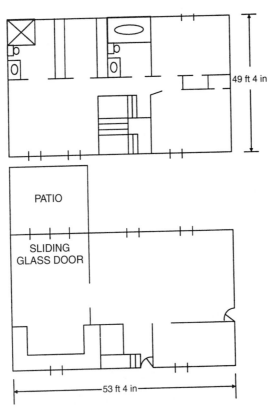

PATIO

SLIDING
GLASS DOOR

49 ft 4 in

53 ft 4 in

22.   Find the cooling load for the house in problem 21. The sun strikes the right side of the house as you face the front door of the house from outside. _____

Note: For problems 23–25, use the following information.

An office building has three floors with similar exterior designs. Each floor is 50 feet by 70 feet with 8½-foot walls. The roof has 9½ inches of insulation. Each floor has 16 windows, each 6 feet high by 3½ feet wide fixed double pane. The first floor is built on a slab and has two doors, each 7 feet high by 4 feet wide instead of two windows. The doors have an insulated core and are weather-stripped. The walls have 3½-inch insulation with 1 inch of polystyrene sheathing. It has been planned for a 75°F design temperature difference.

23.   What is the heat load of the top floor?   _____

24.   What is the heat load of the middle floor?   _____

25.   Find the heat load of the ground floor.   _____

Stretchouts and Lengths of Arcs

Unit 33 STRETCHOUTS OF SQUARE AND RECTANGULAR DUCTS

BASIC PRINCIPLES OF STRETCHOUTS OF SQUARE AND RECTANGULAR DUCTS

• Study and apply these principles of stretchouts of square and rectangular ducts to the problems in this unit.

There are times when a heating and cooling technician must handcraft a duct. It can be done by trial and error or by guesswork, but that can be a waste of both time and material that could be costly. Cutting out the duct correctly the first time is efficient and cost-effective.

In finding the amount of material needed for a square or rectangular duct, the size of the duct pattern, or the *stretchout*, is needed.

Usually, the length of the stretchout (*L.S.*) is the perimeter of the end of the duct plus the allowance for the seam (*M*).

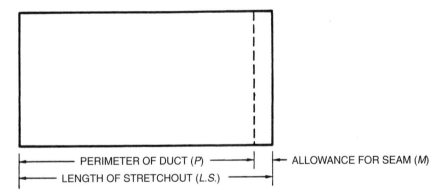

The dimensions of the duct, the width of the seam, and the type of seam will determine the actual layout for the stretchout.

The width of the stretchout (*W.S.*) is usually the length of the duct (*l*) plus the allowance for an overlap for the next duct (*J*), if any is needed.

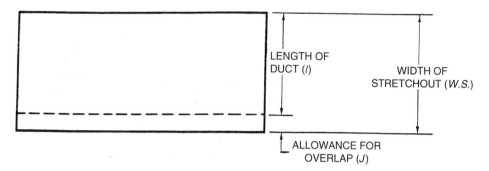

It is important to realize that it is not a requirement to have the length of the stretchout be the perimeter of the duct plus any seam allowance. This could just as easily be the width of the stretchout. Likewise, the width of the stretchout could be the length of the duct plus any overlap. Or this could just as easily be the length of the stretchout. What is important is to realize that any seam allowance goes with the perimeter and any overlap goes with the length of the duct.

Some ducts will have seam allowances on both ends. Some ducts will not have any seam allowance. The actual layout for the stretchout will depend upon the edges, overlaps, and type of seam.

It should be noticed that the seam of the duct runs down the middle of one of the sides of the duct, not at the corner. So, on the stretchout, three sides of the duct will be shown whole and one side will be shown as two halves, one at each end of the stretchout.

Let us take a look at three of the most commonly used types of seams and how the stretchout for each type is determined.

The first type is the *butt* or *welded* seam duct. The two halves of the one side are simply brought together.

BUTT OR WELDED SEAM DUCTS

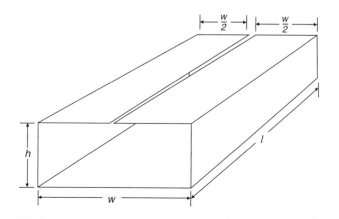

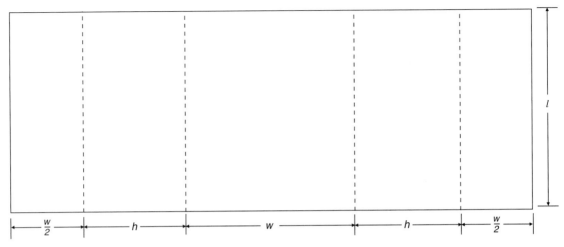

For square ducts: $L.S. = \dfrac{s}{2} + s + s + s + \dfrac{s}{2}$

                         $= 4s$

For rectangular ducts: $L.S. = \dfrac{w}{2} + h + w + h + \dfrac{w}{2}$

                         $= 2h + 2w$

For either duct: $W.S. = l$

                         or $l + J$ if an overlap

The second type is the *lap* seam duct. On this duct, one part of the half-side extends past the other half-side. This makes the lap.

<h3 align="center">LAP SEAM DUCTS</h3>

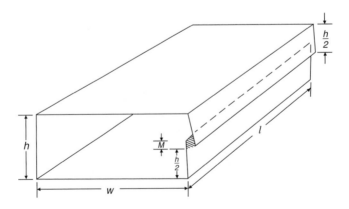

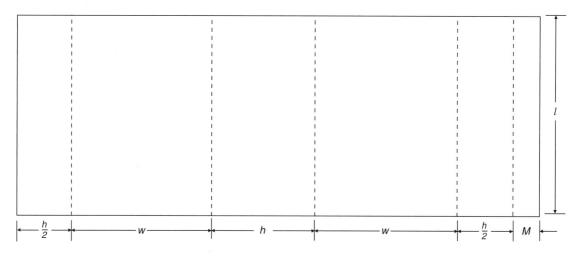

For square ducts: $L.S. = \dfrac{s}{2} + s + s + s + \dfrac{s}{2} + M$

$= 4s + M$

For rectangular ducts: $L.S. = \dfrac{h}{2} + w + h + w + \dfrac{h}{2} + M$

$= 2h + 2w + M$

For either duct: $W.S. = l$

or $l + J$ if an overlap

The third type is the *grooved* seam duct. For this duct, the part extra to the side half is bent in a U or V. The part extra to the other side half is also bent in a U or V. This U or V is hooked into the U or V from the other side half, and then this combination is crimped or welded. The extra portions make a total of three seam widths. In order to have the seam located in the exact middle of the one side, one and one-half of the seam width is added to each end of the stretchout.

GROOVED SEAM DUCTS

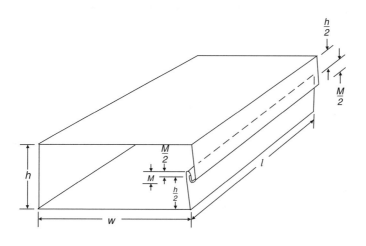

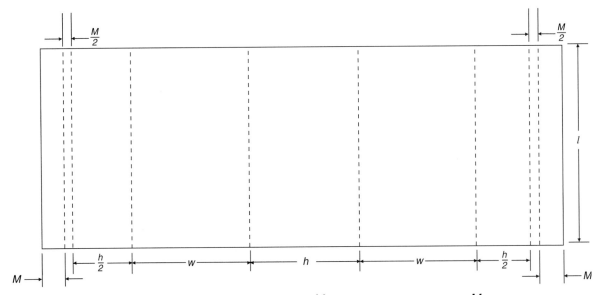

For square ducts:

$$L.S. = M + \frac{M}{2} + \frac{s}{2} + s + s + s + \frac{s}{2} + \frac{M}{2} + M$$

$$= 4s + 3M$$

For rectangular ducts:

$$L.S. = M + \frac{M}{2} + \frac{h}{2} + w + h + w + \frac{h}{2} + \frac{M}{2} + M$$

$$= 2h + 2w + 3M$$

For either duct:

$$W.S. = l$$

or $l + J$ if an overlap

Note: In finding the stretchout for square and rectangular ducts, the thickness of the metal is not considered.

The key to choosing the correct formula is determining the type of seam on the duct.

The length of the seam will always be dimension l plus any overlap.

Review the tables of length equivalents in Section II of the Appendix.

PRACTICAL PROBLEMS

1. A 12-inch square duct is 18 inches long. The duct has a butt seam.

    a. Find the length of the stretchout.

    b. Find the width of the stretchout.

    a. _____

    b. _____

2. A 3-foot-long rectangular duct is 8 inches wide and 10 inches high and has a welded seam.

    a. What is the stretchout length in inches?

    b. What is the stretchout width in inches?

    a. _____

    b. _____

3. The dimensions of a rectangular duct with a butt seam are 6 inches wide by 10 inches high by 26¼ inches long. It has an overlap of ½ inch.

    a. Find the length of the stretchout.

    b. Find the width of the stretchout.

    a. _____

    b. _____

4. A square duct is 27 centimeters on a side and is 1.2 meters long. It has a butt seam.

    a. Find the length of the stretchout.

    b. Find the width of the stretchout.

    a. _____

    b. _____

5. A 3-foot-long square duct has a welded seam and is 9 inches on each side.

    a. Find the length of the stretchout.

    b. Find the width of the stretchout.

    a. _____

    b. _____

6. A rectangular duct is 1 meter long and has a butt seam. The duct is 25 centimeters wide and 20 centimeters high.

    a. Find the length of the stretchout.

    b. Find the width of the stretchout.

    a. _____

    b. _____

7. A 5¾-inch by 12⅛-inch rectangular duct needs to fill a 24-inch gap and needs a 1½-inch overlap to slip into the next piece. The duct has a butt seam.

    a. Find the length of the stretchout.

    b. Find the width of the stretchout.

    a. _____

    b. _____

8. A 2-foot-long rectangular duct is 8 inches wide by 12 inches high. The lap
    seam is ⅜ inch.

    a. Find the length of the stretchout.   a. _____

    b. Find the width of the stretchout.   b. _____

9. The dimensions of a rectangular duct with a lap seam are:

    $h = 25$ cm; $w = 20$ cm; $l = 75$ cm; $M = 0.8$ cm

    a. What is the length of the stretchout in centimeters?   a. _____

    b. What is the width of the stretchout in centimeters?   b. _____

10. An 8-inch square duct has a lap seam of ¼ inch. The duct is 30 inches long.

    a. Find the length of the stretchout.   a. _____

    b. Find the width of the stretchout.   b. _____

11. A rectangular duct is 2 feet wide, 30 inches high, and 2 feet long. The lap
    seam is ¼ inch. The overlap is ¾ inch.

    a. What is the length of the stretchout in inches?   a. _____

    b. What is the width of the stretchout in inches?   b. _____

12. A square duct with a lap seam is to measure 32.4 centimeters on a side. Its
    length is to be 1.2 meters long. The lap is 1.6 centimeters.

    a. Find the length of the stretchout.   a. _____

    b. Find the width of the stretchout.   b. _____

13. A 9⅛-inch square duct has a lap seam of ¾ inches. The duct has a length
    of 3 feet.

    a. Find the length of the stretchout.   a. _____

    b. Find the width of the stretchout.   b. _____

14. A 27½-inch-long rectangular duct is to have a ¾-inch lap seam. The duct is
    6¼ inches by 9⅛ inches.

    a. Find the length of the stretchout.   a. _____

    b. Find the width of the stretchout.   b. _____

15. A rectangular duct is to measure 21.6 centimeters high and 27.3 centimeters wide and have a 2.2-centimeter lap seam. The duct is to be 43 centimeters long.

    a. Find the length of the stretchout.

    b. Find the width of the stretchout.

a. _____

b. _____

16. A square duct is 8 inches on a side. It has a lap seam of 1¼ inches and is 26 inches long with a ¾-inch overlap for the next section.

    a. Find the length of the stretchout.

    b. Find the width of the stretchout.

a. _____

b. _____

17. A square duct measures 22 centimeters on each side. The duct is 1 meter long. It has a 0.75-centimeter grooved seam.

    a. Find in centimeters the length of the stretchout.

    b. Find in centimeters the width of the stretchout.

a. _____

b. _____

18. The dimensions of a square duct with a grooved seam are:

$$s = 9 \text{ inches}; \; l = 3 \text{ feet}; \; M = \tfrac{3}{8} \text{ inch}$$

    a. Find in inches the value of *L.S.*

    b. Find in inches the value of *W.S.*

a. _____

b. _____

19. The width of the grooved seam on a rectangular duct is ¼ inch. The duct is 6 inches wide by 12 inches high by 32½ inches long.

    a. Find the length of the stretchout.

    b. Find the width of the stretchout.

a. _____

b. _____

20. A 30-inch square duct is 27½ inches long. The grooved seam is ⅜ inch. There is a ¼-inch overlap on both ends.

    a. Find the total length of the stretchout.

    b. Find the total width of the stretchout.

a. _____

b. _____

21. A rectangular duct is 1½ feet wide by 30 inches high by 22 inches long. It has a ¼-inch grooved seam. The overlap is 1 inch.

    a. What is the value of *L.S.*?

    b. What is the value of *W.S.*?

a. _____

b. _____

22. A square duct 25.3 centimeters on a side and 75 centimeters long has a 1.2-centimeter grooved seam.

   a. Find the length of the stretchout.

   b. Find the width of the stretchout.

a. _____

b. _____

23. A square duct measures $10\frac{5}{8}$ inches on a side with a length of 32 inches. It is to have a $\frac{3}{4}$-inch grooved seam.

   a. Find the length of the stretchout.

   b. Find the width of the stretchout.

a. _____

b. _____

24. A duct is a rectangle $9\frac{1}{8}$ inches high and $13\frac{3}{4}$ inches wide. The duct is 30 inches long with a $\frac{3}{4}$-inch grooved seam.

   a. Find the length of the stretchout.

   b. Find the width of the stretchout.

a. _____

b. _____

25. A rectangular duct is 24.7 centimeters high and 32.6 centimeters wide. It has a grooved seam 1.1 centimeters wide. This duct is to be 78 centimeters long and have an 8-centimeter overlap.

   a. Find the length of the stretchout.

   b. Find the width of the stretchout.

a. _____

b. _____

Unit 34 STRETCHOUTS OF CIRCULAR DUCTS

BASIC PRINCIPLES OF STRETCHOUTS OF CIRCULAR DUCTS

- Study and apply these principles of stretchouts of circular ducts to the problems in this unit.

In finding the amount of material needed for a circular duct, the size of the duct pattern, or the *stretchout*, is needed.

Determining the stretchout for a circular duct is similar to finding the stretchout for a rectangular duct, except that a different formula is used. Up to this point, the circumference of a circle has been found using the following formula:

$$C = 2 \times \pi \times r$$

If the diameter of the circle was given, the radius could be found using this formula:

$$r = \frac{D}{2}$$

There is a second formula that can be used to find the circumference:

$$C = \pi \times D$$

Most circular ducts are sized giving the diameter. This formula makes finding the circumference easier when the diameter is given.

The length of the stretchout (*L.S.*) is the circumference at the end of the duct (*C*) plus the allowance for the seam (*M*). The width of the stretchout (*W.S.*) for circular ducts without overlap is the length of the duct (*l*). Just like in the last unit, the length of the stretchout and the width of the stretchout could be interchanged.

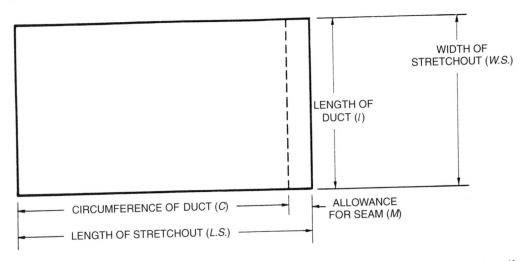

The dimensions of the duct, the width of the seam, the type of seam, and the overlap, if any, will determine the actual layout for the stretchout.

Just as with rectangular ducts, there are three more commonly used types of seams. Let us look at them and the formulas that will determine the stretchout for each type.

The first type is the *butt* or *welded* seam duct.

BUTT OR WELDED SEAM DUCTS

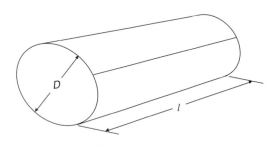
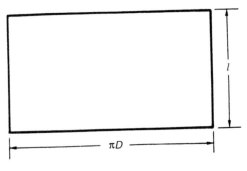

For circular ducts:
$$L.S. = C$$
$$= 2\pi r$$
$$\text{or } = \pi D$$
$$\text{use } \pi = 3.1416$$
$$W.S. = l$$
$$\text{or } = l + J \text{ if an overlap}$$

The second type is the *lap* seam duct. This is lapped in the same manner as the rectangular duct.

LAP SEAM DUCTS

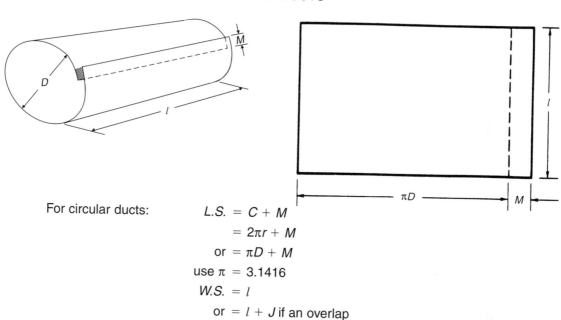

For circular ducts:

$$L.S. = C + M$$
$$= 2\pi r + M$$
$$\text{or} = \pi D + M$$
$$\text{use } \pi = 3.1416$$
$$W.S. = l$$
$$\text{or} = l + J \text{ if an overlap}$$

The third type is the *grooved* seam duct. Its design is just as described in the previous unit. The difference here is that there is no middle of a particular side, so we do not have to split one seam width and add it to each end of the length of the stretchout.

GROOVED SEAM DUCTS

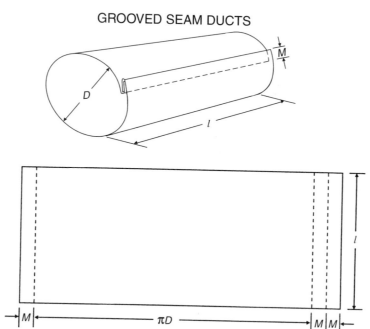

For circular ducts: $L.S. = C + M + M + M$
                            or $= \pi D + 3M$
                            use $\pi = 3.1416$
                            $W.S. = l$
                            or $= l + J$ if an overlap

Note: In finding the stretchout for circular ducts made with gauge metal, the thickness of the metal is not considered.

The key to choosing the correct formula is determining the type of seam on the duct.

The length of the seam will always be dimension l plus any overlap.

Review the tables of length equivalents in Section II of the Appendix.

Example: Find the length to the nearer hundredth centimeter of the stretchout and the width of the stretchout for a 2-meter-long circular duct 10 centimeters in radius with a 1.3-centimeter lap seam and a 2.5-centimeter overlap.

$r = 10$ cm $M = 1.3$ cm
$l = 2$ m $= 200$ cm $J = 2.5$ cm

For a lap seam:
$L.S. = 2\pi r + M$
$= 2\,(3.1416)\,10 + 1.3$ cm

Estimating:
$= 2(3)\,10 + 1.3$
$= 60 + 1.3$
$= 61.3$ cm

Calculating:
$= 62.832 + 1.3$
$= 64.13$ (rounded off)

$W.S. = l + J$
$= 200 + 2.5$
$= 202.5$ cm

PRACTICAL PROBLEMS

1. The diameter of a circular duct is 6 inches. The duct is 24 inches long and has a butt seam.

    a. Find to the nearer 16th inch the length of the stretchout.

    a. _____

    b. Find the width of the stretchout.

    b. _____

2. A circular duct is 20 inches long with a butt seam. The diameter of the duct is 8½ inches.

    a. What is the length of the stretchout? Round the answer to the nearer 16th inch.

    a. _____

    b. What is the width of the stretchout?

    b. _____

3. A 24-centimeter diameter duct is 50 centimeters long and has a butt seam.

    a. Find to the nearer hundredth centimeter the value of *L.S.*

    a. _____

    b. Find in centimeters the value of *W.S.*

    b. _____

4. A duct has a butt seam and is 3 feet long. It has a radius of 5⅛ inches.

    a. Find to the nearer 16th inch the length of the stretchout.

    a. _____

    b. Find the width of the stretchout.

    b. _____

5. A circular duct has a radius of 16.3 centimeters. It is 1.1 meters long and has a welded duct.

    a. Find to the nearer hundredth centimeter the length of the stretchout.

    a. _____

    b. Find the width of the stretchout.

    b. _____

6. A circular duct with a butt seam has a radius of 4 inches and a length of 28 inches and is to have an overlap of 3¼ inches.

    a. Find to the nearer 16th inch the length of the stretchout.

    a. _____

    b. Find the width of the stretchout.

    b. _____

7. A circular duct is to measure 20.5 centimeters in diameter and 75 centimeters long. It has a butt seam.

    a. Find to the nearer hundredth centimeter the length of the stretchout.

    a. _____

    b. Find the width of the stretchout.

    b. _____

8. A stretchout for a circular duct with a welded seam is needed. The duct is to measure 14 centimeters in radius and 0.9 meter long.

    a. Find to the nearer hundredth centimeter the length of the stretchout.   a. _____

    b. Find the width of the stretchout.   b. _____

9. The dimensions of a circular duct with a lap seam are:
    $$D = 8 \text{ inches}; \ l = 3 \text{ feet}; \ M = \tfrac{1}{4} \text{ inch}$$

    a. What is the value of *L.S.* to the nearer 16th inch?   a. _____

    b. What is the value of *W.S.* in inches?   b. _____

10. The width of the lap seam on a circular duct is 0.7 centimeter. The duct is 1 meter long and has a diameter of 30 centimeters.

    a. Find to the nearer hundredth centimeter the length of the stretchout.   a. _____

    b. Find in centimeters the width of the stretchout.   b. _____

11. A 10-inch diameter duct has a length of $2\tfrac{1}{2}$ feet. The lap seam has a width of $\tfrac{3}{8}$ inch.

    a. What is the length of the stretchout? Round the answer to the nearer 16th inch.   a. _____

    b. What is the width of the stretchout?   b. _____

12. A circular duct with a 1 centimeter lap seam is to measure 27.5 centimeters in diameter and 95 centimeters long.

    a. Find to the nearer hundredth centimeter the length of the stretchout.   a. _____

    b. Find the width of the stretchout.   b. _____

13. A duct with a radius of $6\tfrac{1}{4}$ inches is 3 feet long and has a $\tfrac{3}{4}$-inch lap seam.

    a. Find to the nearer 16th inch the length of the stretchout.   a. _____

    b. Find the width of the stretchout.   b. _____

14. A circular duct is to have the following dimensions: radius, 12.2 centimeters; length, 1.2 meters; and seam width, 0.8 centimeter. The duct needs a 2.2-centimeter overlap.

    a. Find to the nearer hundredth centimeter the length of the stretchout.   a. _____

    b. Find the width of the stretchout.   b. _____

15. A circular duct measures 5½ inches in radius and is 30 inches long. It has a ⅝-inch lap seam and a ⅞-inch overlap.

    a. Find to the nearer 16th inch the length of the stretchout.

    b. Find the width of the stretchout.

        a. _____

        b. _____

16. The radius of a circular duct is 14 centimeters. The duct is 70 centimeters long and has a 0.9-centimeter lap seam.

    a. Find to the nearer hundredth centimeter the length of the stretchout.

    b. Find the width of the stretchout.

        a. _____

        b. _____

17. The diameter of a circular duct is 7½ inches. The length of the duct is 28 inches. The grooved seam on the duct has a width of ⅜ inch.

    a. How many inches long is the stretchout? Round the answer to the nearer 16th inch.

    b. How many inches wide is the stretchout?

        a. _____

        b. _____

18. A 9-inch diameter duct has a ¼-inch-wide grooved seam. The length of the duct is 2 feet.

    a. Find to the nearer 16th inch the length of the stretchout.

    b. Find the width of the stretchout.

        a. _____

        b. _____

19. The dimensions of a circular duct with a grooved seam are:

    D = 22 centimeters; l = 70 centimeters; M = 0.8 centimeter

    a. What is the value of *L.S.* to the nearer hundredth centimeter?

    b. What is the value of *W.S.* in centimeters?

        a. _____

        b. _____

20. A duct with a grooved seam measures 6¼ inches in diameter and is 32 inches long. It has a seam width of ¾ inch.

    a. Find to the nearer 16th inch the length of the stretchout.

    b. Find the width of the stretchout.

        a. _____

        b. _____

21. A circular duct is to be 20.2 centimeters in diameter and 90 centimeters long with a 1 centimeter grooved seam.

    a. Find to the nearer hundredth centimeter the length of the stretchout.

    b. Find the width of the stretchout.

        a. _____

        b. _____

22. A circular duct is to have the following dimensions: radius, 11.4 centimeters; length, 0.75 meter; grooved seam width, 0.9 centimeter; and overlap, 1.8 centimeters.

    a. Find to the nearer hundredth centimeter the length of the stretchout.   a. _____

    b. Find the width of the stretchout.   b. _____

23. A circular duct measures 3 feet long with a radius of 4 inches. It has a $\frac{5}{8}$-inch grooved seam.

    a. Find to the nearer 16th inch the length of the stretchout.   a. _____

    b. Find the width of the stretchout.   b. _____

24. A grooved seam, circular duct is $4\frac{1}{2}$ inches in radius, 34 inches long, with a $\frac{3}{4}$-inch wide seam and an overlap of $1\frac{1}{2}$ inches.

    a. Find to the nearer 16th inch the length of the stretchout.   a. _____

    b. Find the width of the stretchout.   b. _____

25. A duct is 1.2 meters long and 10.5 centimeters in radius. It has a 1.2-centimeter grooved seam.

    a. Find to the nearer hundredth centimeter the length of the stretchout.   a. _____

    b. Find the width of the stretchout.   b. _____

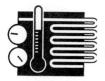 ## Unit 35 *LENGTHS OF ARCS OF CIRCLES*

BASIC PRINCIPLES OF LENGTHS OF ARCS OF CIRCLES

- Study and apply these principles of lengths of arcs of circles to the problems in this unit.

There are times when ducts must make turns. These ducts will have to be made from separate pieces. In order to properly cut out the pieces for the ducts, we need to study and become familiar with arcs of circles.

An arc is part of a circle. The number of degrees in an arc is measured by the central angle. The length of an arc or arc length (L) is a fraction times the circumference (C) of the circle. The fraction is a ratio of the number of degrees in the arc ($n°$) to the number of degrees in the circle ($360°$).

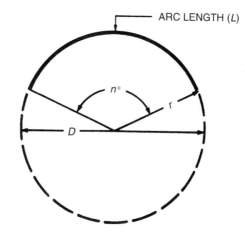

$$L = \frac{n°}{360°} \times \pi D$$

$$L = \frac{n°}{360°} \times 2\pi r$$

use $\pi = 3.1416$

For a given arc length and circumference, the measure of the angle is:

$$n° = \frac{360° \times L}{\pi D}$$

or

$$n° = \frac{360° \times L}{2\pi r}$$

use $\pi = 3.1416$

A *rectangular 90° elbow* is an elbow with the heel and the throat at 90° to the cheek, and the cheek is a 90° portion of a circle. A *rectangular 45° elbow* is an elbow with the heel and the throat at 90° to the cheek, and the cheek is a 45° portion of a circle.

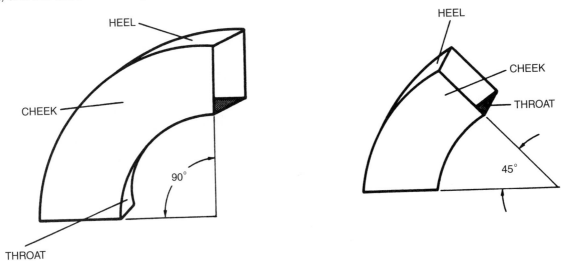

To find the length of the stretchout of the throat or the heel, the arc length is needed. Arc length is found by using the throat radius or the heel radius. The heel radius is the throat radius plus the width of the duct.

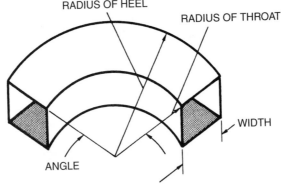

Note: The radius of the heel of a rectangular elbow is not the width of the duct, but the distance from the center of the arc.

For a circular duct, the outside radius is the inside radius plus the diameter of the duct.

If you know the arc length and the number of degrees in the arc, the diameter of the circle out to that arc length can be found by first solving the arc length formula for *D*.

$$L = \frac{n°}{360°} \times \pi D$$

$$\pi D = \frac{L \times 360°}{n°}$$

$$D = \frac{L \times 360°}{\pi \times n°}$$

Example: A 40° arc has a radius of 2½ inches. Find the length of the arc.

$$L = \frac{n°}{360°} \times 2\pi r$$

$$= \frac{40°}{360°} \times 2(3.1416)2.5$$

Estimating: $= \dfrac{40}{400} \times 2(3)3$

$= \dfrac{1}{10} \times 18$

$= 1.8$ in

Calculating: $= 1.745$ in (rounded off to the nearer thousandth)

PRACTICAL PROBLEMS

1. Find to the nearer hundredth inch the length of this arc.   _____

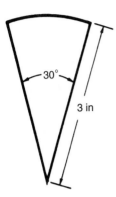

2. What is the length of this arc in centimeters? Round the answer to the nearer hundredth.   _____

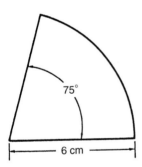

3. This arc has a radius of 4 feet. The central angle is 225°. How many feet are in the measure of the arc? Round the answer to the nearer hundredth.   _____

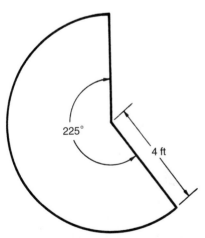

4.   A 270° arc has a radius of 0.5 meter. Find to the nearer hundredth meter the length of the arc.

_____

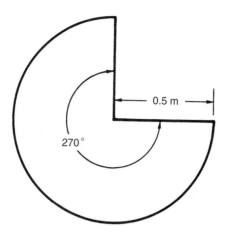

0.5 m

270°

5.   The length of an arc of a circle is 11.775 feet. The diameter of the circle is 9 feet. How many degrees are in the central angle of the arc? Round the answer to the nearer degree.

_____

6.   An arc has a length of 86.55 centimeters. The radius is 15 centimeters. Find the number of degrees in the central angle of the arc. Round the answer to the nearer degree.

_____

7.   The cylinder of a rotary compressor is 12 centimeters in diameter. The angle between the intake and exhaust ports of the compressor is 40°. What is the distance between the ports measured along the arc? Round the answer to the nearer hundredth centimeter.

_____

8.   A metal screw on the knob of the thermostat dial turns to adjust the bimetallic strip so that the thermostat will turn on and off at different temperatures. The screw is located 1.75 centimeters from the center of the knob. The knob turns through 195° going from OFF to its highest setting. How far does the screw travel when going from OFF to the highest setting? Round to the nearer hundredth centimeter.

_____

9.   What is the distance along the outside edge?   _____

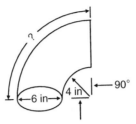

10.   What is the arc length of the center of this duct? Round the answer to the
       nearer hundredth inch.   _____

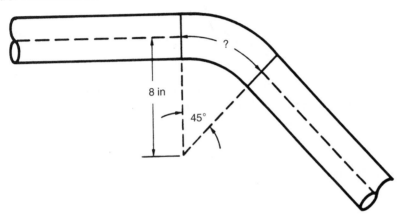

11.   Openings in a gas furnace allow air to enter the burners. The openings of this
       furnace have a central arc length of ¾ inch. Find to the nearer degree the
       angle of one of the openings.   _____

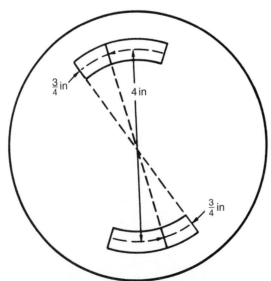

12. The control knob of a window air-conditioning unit has four settings. The knob is turned from the OFF position to the HI COOL position. Find the number of degrees through which the knob is turned. Round the answer to the nearer whole degree.

_____

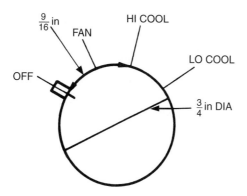

13. An oil gun is fastened to a furnace with six screws. The screws are equally spaced and form a 6-inch diameter circle. What is the arc length to the nearer hundredth inch between the centers of the screws?

_____

14. There are nine equally spaced markings along the edge of the circular temperature control dial of a refrigerator. The arc length between the markings is 2.7 centimeters. Find the diameter of the dial. Round the answer to the nearer hundredth centimeter.

_____

15. The rectangular 45° elbow has a throat radius of 5 inches. The heel radius is 15 inches.

   a. What is the arc length of the throat to the nearer 16th inch?

_____

   b. What is the arc length of the heel to the nearer 16th inch?

_____

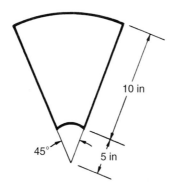

16. The throat radius of a rectangular 90° elbow is 5 inches. The duct is 10 inches wide.

   a. Find to the nearer 16th inch the length of the throat.   a. _____

   b. Find to the nearer 16th inch the length of the heel.   b. _____

17. A rectangular 45° duct is 9 inches wide. The radius of the throat is 4½ inches.

   a. What is the arc length of the throat to the nearer 16th inch?   a. _____

   b. What is the arc length of the heel to the nearer 16th inch?   b. _____

18. A 30-centimeter-wide duct has a throat radius of 15 centimeters. The angle between the sides of the heel and the cheek is 90°.

   a. Find to the nearer hundredth centimeter the arc length of the throat.   a. _____

   b. Find to the nearer hundredth centimeter the arc length of the heel.   b. _____

19. A 25.6-centimeter-wide duct has a heel radius of 35 centimeters. This duct is a 90° elbow.

   a. Find the arc length of the heel.   a. _____

   b. Find the arc length of the throat.   b. _____

20. A circular 7-inch duct has an inside radius of 6 inches. This duct is a 45° elbow.

   a. Find the inside arc length.   a. _____

   b. Find the outside arc length.   b. _____

21. A 90° elbow of a circular 12-centimeter diameter duct has an inside radius of 10 centimeters.

   a. Find the inside arc length.   a. _____

   b. Find the outside arc length.   b. _____

22. A rectangular duct has a throat radius of 6 inches. The duct is 6½ inches wide. This duct is a 90° elbow.

   a. Find the inside arc length.   a. _____

   b. Find the outside arc length.   b. _____

23. A 45° elbow of a rectangular duct has a 12.4-centimeter throat radius. The duct width is 20.5 centimeters.

   a. Find the inside arc length.   a. _____

   b. Find the outside arc length.   b. _____

24. A circular 9-inch duct has an inside radius of 6½ inches. The duct is a 90° elbow.

   a. Find the inside arc length.

   b. Find the outside arc length.

   a. _____

   b. _____

25. A rectangular duct is 8¼ inches wide. The throat radius of this 45° elbow is 5½ inches.

   a. Find the inside arc length.

   b. Find the outside arc length.

   a. _____

   b. _____

Trigonometry

Unit 36 TRIGONOMETRIC FUNCTIONS

BASIC PRINCIPLES OF TRIGONOMETRIC FUNCTIONS

- Study and apply these principles of trigonometric functions to the problems in this unit.

- Use the table of values for trigonometric functions found in Section VI of the Appendix.

Triangles have the following property: when two triangles have the same angles, the length of their sides will have the same ratios. This property can be used to determine angles or lengths of sides of triangles. The ratios of the length of the sides of a right triangle—a triangle with one angle equal to 90°—are given special names and put in a table.

These three relationships are widely used in expressing the ratio of the sides of a right triangle. These relationships are called trigonometric functions.

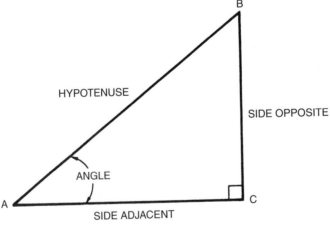

The trigonometric functions and their ratios are:

$$\text{SINE} \quad \sin = \frac{\text{side opposite}}{\text{hypotenuse}} \qquad \text{COSINE} \quad \cos = \frac{\text{side adjacent}}{\text{hypotenuse}}$$

$$\text{TANGENT} \quad \tan = \frac{\text{side opposite}}{\text{side adjacent}}$$

It is important to realize that whether a side is opposite or adjacent depends upon which angle you are working with. When a right triangle is drawn, the right angle is always labeled as angle C.

Two lines—two sides of the triangle—come together to make each angle of the triangle. One side of the triangle is not involved in making the angle. This side is designated the opposite side for that angle. For each angle, a different side of the triangle is designated as the opposite side. The side that is opposite the right angle is always designated as the hypotenuse. As can be seen in the following figure, the adjacent and opposite labels are different for the different angles.

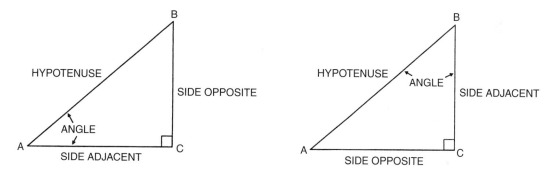

In either case, the definitions for the ratios stay the same. In other words, the sine is the opposite side length divided by the hypotenuse no matter which angle you are talking about. The opposite side is different for the different angles, so the value of the sine will be different for different angles.

These ratios and the table of trigonometric functions found in the Appendix are used to find the sides and angles of a right triangle.

The formula for a trigonometric function can be rearranged to find any one of the three values that make up the formula. The formula for sine (sin) is:

$$\sin = \frac{\text{side opposite}}{\text{hypotenuse}}$$

This is the formula we would use if we are trying to find the value for the sine. If we know the sine and are trying to find the value for the length of one of the sides, we would use

$$\text{side opposite} = \sin \times \text{hypotenuse}$$

or

$$\text{hypotenuse} = \frac{\text{side opposite}}{\sin}$$

Similar formulas would be used to find values using the other trigonometric functions.

When solving problems, draw a right triangle and label the parts (hypotenuse, opposite, A, and so on). Then put the values that are given in the problem on the triangle. This will give a better picture of what you are looking for, what trigonometric function will give it to you, and what formula to use to find the answer.

Example 1: Find the size of angle A to the nearer whole degree.

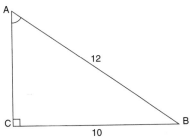

For angle A, 10 is the opposite side and 12 is the hypotenuse. The trigonometric function that relates these two sides is the sine.

$$\text{Sin } A = \frac{\text{opposite side}}{\text{hypotenuse}} = \frac{10}{12} = 0.8333$$

From the Table of Trigonometric Functions in the Appendix, the angle whose sine is closest to 0.8333 is 56°. So angle A is 56°.

Example 2: Find the length of side a.

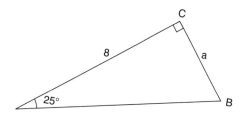

For angle A, a is the opposite side and 8 is the adjacent side. The trigonometric function that relates the opposite and adjacent sides is the tangent.

$$\text{Tan } A = \frac{\text{opposite side}}{\text{adjacent side}}$$

$$\text{Tan } 25° = \frac{a}{8}$$

From the Table of Trigonometric Functions in the Appendix, Tan 25° = 0.4663.

$$0.4663 = \frac{a}{8}$$

$$a = 8 \times 0.4663$$

$$= 3.7304$$

PRACTICAL PROBLEMS

Note: For problems 1–8, round the answer to the nearer whole degree.

1.   Find the size of angle A.

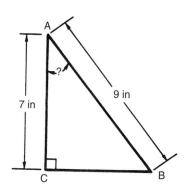

2.   How many degrees are there in angle E?

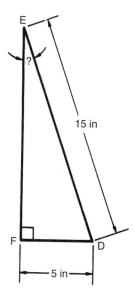

3.   Determine the value of angle Q.

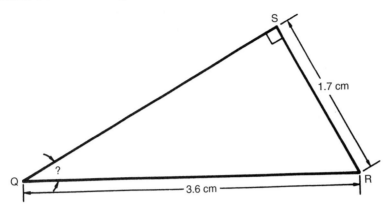

4.   In triangle XYZ, side XZ is 1½ feet long and side YZ is 2¼ feet long.

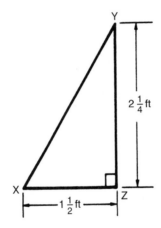

a. Find the size of angle X.

b. Find the size of angle Y.

5. The cover of a baseboard heater has panels to deflect the air into the heater. Through what angle has the panel been bent when the cover is built?

_____

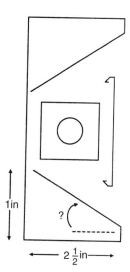

6. A microswitch on a thermostat controls the heater in that room. The bimetallic strip is ⅞ inch long and the end of the microswitch moves ¹⁄₃₂ inch from ON to OFF. What angle does the contact pass through going from ON to OFF?

_____

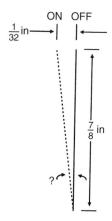

7. To reduce heat buildup in the summer in an attic, a ridge vent is installed on the roof of a house. The base of the vent should have the same angle as the roof.

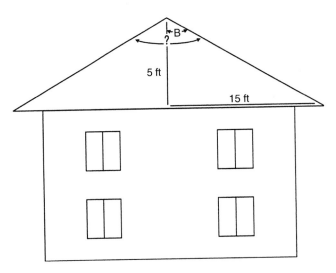

a. Find the value of angle B.

b. What angle should the vent have? (Find angle ?.)

8. A duct forms a Y. Find the angle that the two arms of the Y make.

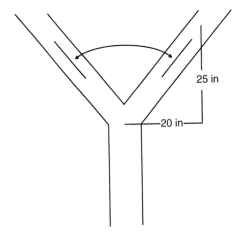

Note: For problems 9–12, round the answer to the nearer hundredth.

9. How many centimeters long is side AC?   _____

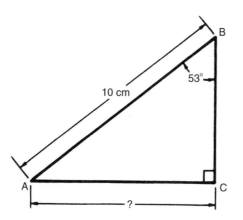

10. Find the length of side EF.   _____

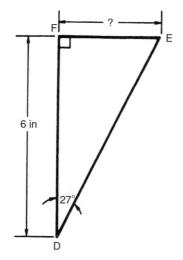

11. What is the length of the hypotenuse in triangle MNO?   _____

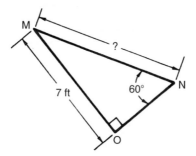

12.   Find the length of side TS.

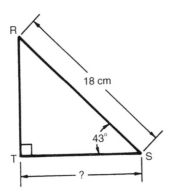

13.   A fuel line for an oil burner is installed in a straight line from the tank to the burner. What is the vertical drop in the fuel line? Round the answer to the nearer hundredth foot.

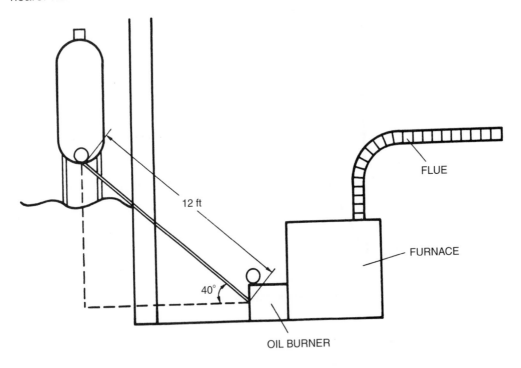

14. A round trunk duct fitting has a branch that forms an angle of 30° with the main duct. The main duct is straight. The main duct and the branch form a Y. How far apart are the centers of the ducts 25 feet from the Y along the center of the branch? Round the answer to the nearer hundredth foot.

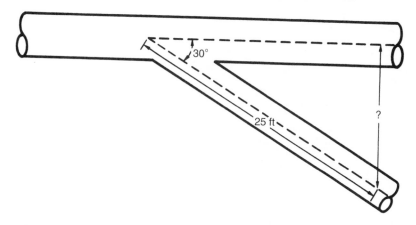

15. A section of duct is used to reduce the height of the duct. Find to the nearer degree the angle formed by the side of this reducer.

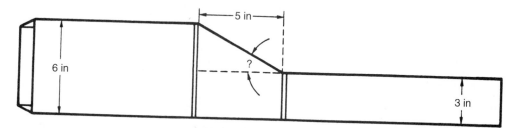

16. In an ice maker, ice cubes fall from a tray into a bin. The tray is 25 inches long. The front of the tray is 10 inches lower than the back of the tray. At what angle is the tray placed? Round the answer to the nearer whole degree.

17. What is the distance along the grill of this baseboard spreader? Round the
     answer to the nearer hundredth inch.   _____

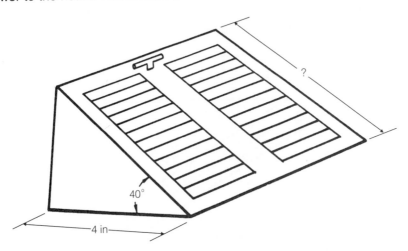

18. A compressor reed valve is made from a flat piece of metal. The reed is 0.8
     centimeter long. When gas passes through the port, the movable end of the
     reed makes an angle of 25° with the normal position. How high is the movable
     end of the reed when the valve is open? Round the answer to the nearer
     hundredth centimeter.   _____

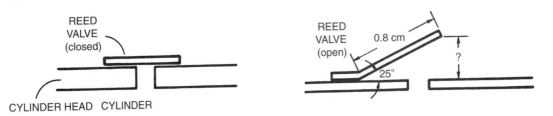

19. Find to the nearer hundredth inch the length of the side of the reducer of this
     duct.   _____

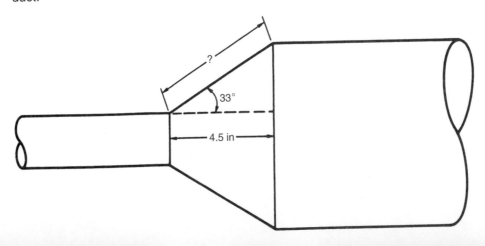

20. A support is built for a duct. Find to the nearer degree the angle that the support makes with the duct.

_____

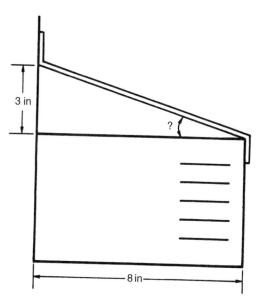

21. The Good Eating restaurant is adding suspended outdoor radiant heating to its 25 feet × 60 feet patio to increase dining space. The duct is to run diagonally from one corner of the patio to the other. It starts and ends 18 inches from each corner. The burner takes up 4 feet of the length. What is the length of the remaining heater duct?

_____

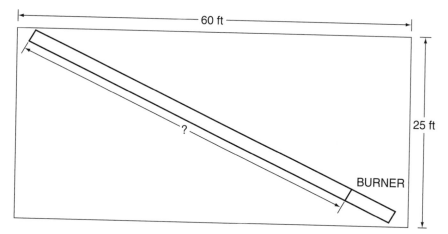

22. An oil burner nozzle is labeled 60°. This is the angle from straight ahead to the direction that the oil is sprayed. A piece of paper is placed 12 inches from the nozzle. How far away from the straight-ahead mark does the oil strike the paper? Round the answer to the nearer hundredth inch.

_____

23. A flue pipe joins the opening of an oil burner to the opening of a chimney. The chimney is 12 feet away from the oil burner. The opening on the chimney is 3 feet higher than the opening on the burner.

   a. What angle does the bottom of the flue pipe make with the horizontal? Round the answer to the nearer whole degree.

   a. _____

   b. Find to the nearer hundredth foot the length of the flue pipe between the chimney and the burner.

   b. _____

24. Each vane of a diffuser is bent at an angle of 30°. Each vane is 1 centimeter long. How far will the edge of the vane stick out from the wall?

_____

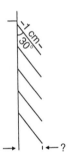

25. A solar heating panel is placed 43° from the vertical. The panel is 8 feet high. What is the length and height for a frame for a flat roof?

length _____

height _____

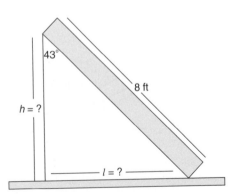

Graphs

SECTION 11

Unit 37 GRAPHS AND GRAPHING

BASIC PRINCIPLES OF GRAPHS AND GRAPHING

- Review and apply the principles of graphing to the problems in this unit.

Graphs are ways of easily displaying quantities of information. A single graph can display the relationship between two variables. Those two variables are given on the two axes of the graph. Each axis should be a uniform scale, but it is important to realize that the two axes do not have to have the same scale. Each axis should have the units and the scale displayed on it.

Additional information can be displayed on a single graph, but additional axes are needed, or additional lines (curves) must be put on the graph. In this case, care must be taken to follow the correct line to the correct axis to obtain the value that you are trying to find.

Usually, you will use a graph that has already been drawn to get information. There may be some times when you want to use a graph to display information better than just listing it in a table. So, there may be times when you want to draw a graph.

When drawing a graph, you must decide what you want to display. Then you have to determine the extent of your values. The whole graph should be filled with the values that are being displayed, not just a small portion of the graph.

Note: A graph may have many variables displayed on it at the same time. When reading such a graph, be certain that you are following the proper line to the correct scale to determine what the value is. A ruler or straight edge often helps when reading the graph, but be very careful using a straight edge when reading a chart that has curved lines.

The scale should tell you the units of the graph. Make sure that you include all of that information when you are drawing a graph. You also want to make sure that your data fills the graph, not just a very tiny portion of it, which would make the reading of your graph difficult.

There are different types of graphs. They display the data in different ways. The most often used type of graph is the line graph where the data is shown as a line. The line can be a straight line or a curved line, but a line connects the different data points. A different type of graph is the bar graph. On this type of graph, the data is displayed as a series of bars. The length or height of each bar is related to the size of the quantity being displayed by the graph. Each bar represents a different data point.

PRACTICAL PROBLEMS

Note: Problems 1–3 deal with the information on the following graph, which shows the number of calls made by the Warm and Cozy Heating Company for a year.

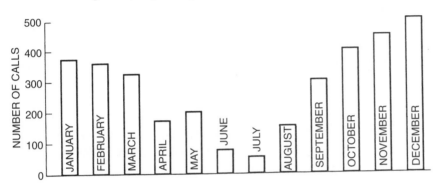

1. When would be the best time to overhaul the service trucks?   _____

2. An average technician can make 3 calls in an 8-hour day.

   a. How many calls would a technician make in a month? (An average work month has 22 days in it.)   a. _____

   b. What is the largest number of trucks (and technicians) that are needed by the Warm and Cozy Heating Company?   b. _____

3. It has been found that during June and July, running a "Heating System Tune-up Special" will increase the number of calls made by 250%. If that special were run this year, how many technicians would be needed during the month of June?   _____

Note: Problems 4 and 5 refer to the following graph for the Keep Kool Company.

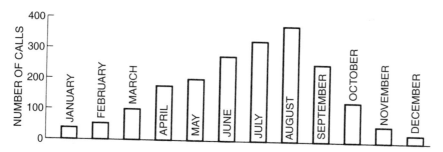

4. How many repair calls were made in the month of September?

_____

5. An average technician makes 3 calls in an 8-hour day.

   a. What is the greatest number of technicians needed?

   a. _____

   b. What is the smallest number of technicians needed?

   b. _____

   Note: It is decided to merge the Warm and Cozy Heating Company and the Keep Kool Company to form the Hot and Cold Service Company. Use this information and the graphs used in the previous problems for problems 6–9.

6. Make a bar graph of the combined number of service calls on a monthly basis for the Hot and Cold Service Company.

7. Make a line graph of the information in problem 6.

8. If part-time technicians can be hired, what is the number of technicians that would be kept busy full-time?

_____

9. What is the largest number of part-time technicians needed to help the full-time technicians?

_____

Note: Problems 10–13 deal with the following simplified pressure heat diagram for R-134a refrigerant.

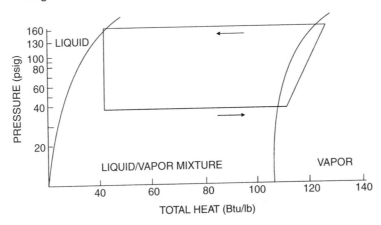

The straight lines on the graph represent what happens to the refrigerant during one cycle of the refrigeration cycle. At the lower right corner, refrigerant vapor enters the compressor. The pressure of the gas is increased, and its temperature (or total heat) is also increased. The hot gas exits the compressor and goes to the cooling (condenser) coils. Passing through the cooling coils allows the gas to give up some of its heat. In doing that, the refrigerant begins to condense. More and more of the gas condenses until we have a pure liquid, as shown in the upper left part of the graph. This liquid then passes through an expansion valve where the pressure on the liquid is decreased. Some of the liquid boils and becomes vapor. The liquid-vapor mixture moves to the evaporator where the liquid boils, becoming all vapor. In this last step, the heat to boil the liquid comes from the area that is to be cooled. We now have all vapor that enters the compressor ready to start the process again.

10.   A refrigerator uses 8 ounces in its system. How much heat is given up by the R-134a when all of it is cooled from a compressed vapor to a liquid?   _____

11.   How much heat does the full 8 ounces remove from the refrigerator area as it boils back to a vapor?   _____

12.   How many times must the 8-ounce charge of refrigerant go through the cycle to remove the heat put into the refrigerator when hot gelatin is put into the refrigerator to cool? The 4 pounds of liquid go from 180°F (720 Btu) to 60°F (240 Btu). Round the answer to the nearer tenth.   _____

13.   How much heat is added by the pump to 1 pound of the refrigerant and must be removed without doing any cooling?   _____

Note: Problems 14–16 deal with the following chart for a heat pump system.

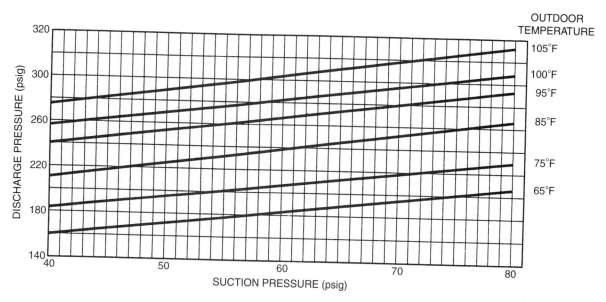

14. If the compressor suction pressure is 70 pounds per square inch gauge pressure (psig) and the outside temperature is 85°F, what should the discharge pressure be reading?

_____

15. If a system is overcharged, the discharge pressure will read higher than it should. If the system is undercharged, the discharge pressure will read lower than it should. The compressor suction pressure is 62 psig and the discharge pressure indicates 290 psig when the outside temperature is 85°F. Should the refrigerant be added to the system or taken out?

_____

16. A correctly charged system is checked in the morning when the outdoor temperature is 75°F. The suction pressure indicates 58 psig. A second check of the system is made on an afternoon when the outdoor temperature is 90°F. There is no change in the suction pressure reading. What should the old and new discharge pressure be?

_____

Note: Problems 17–19 deal with the abbreviated psychrometric chart below.

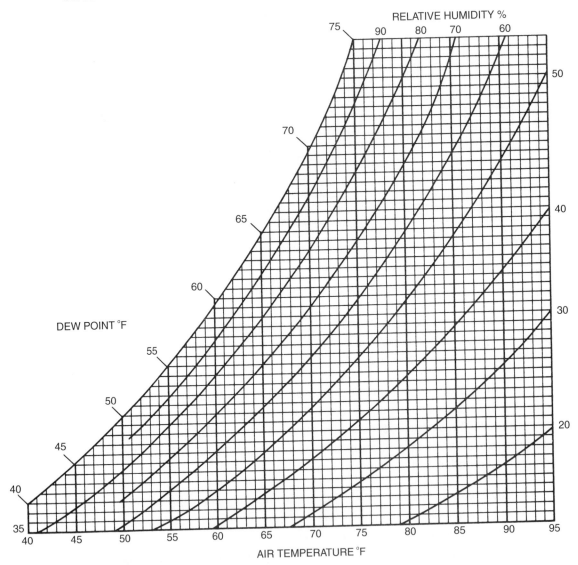

The dew point is the temperature of a quantity of air where it cannot hold any more water vapor. Relative humidity is the amount of water vapor in the air compared to the maximum amount the air could hold at that temperature. Heating the air up without adding moisture (water vapor) will lower the relative humidity; cooling the air down will raise the relative humidity.

If air is cooled below its dew point, moisture will come out of the air. This is what happens on glasses of cold drinks or cold pipes in the summer. It can also happen to air-conditioning ducts running through warmer rooms or air spaces. The psychrometric chart gives information about the dew point and relative humidity.

The chart is read by locating the air temperature along the bottom scale (horizontal axis) and then moving vertically upward until the relative humidity is reached (curved lines). Moving horizontally to the left from that point to the edge of the chart (100% humidity) will give the dew point.

17. A duct carrying 55°F air passes through air spaces listed below. In each case, determine whether the duct will sweat (have moisture form on it).

   a. The room has 70°F air with a relative humidity of 50%.   a. _____

   b. The room has 85°F air with a relative humidity of 50%.   b. _____

   c. The room has 90°F air with a relative humidity of 35%.   c. _____

18. If 65°F air with a 60% relative humidity is heated to 85°F, what is its relative humidity at this new temperature?

   _____

19. Below what humidity for 95°F air will we not have to worry about sweating as the air flows past 72°F pipes?

   _____

Note: Problems 20–22 deal with a pump in a system that is represented by the head/flow curve below. The pump curve shows how the discharge pressure from the pump (pump head) is related to the flow rate. The system curve shows how discharge pressure from a pump will force water through the pipes and valves that make up a flow system. Where these two curves cross is where the pump in the system will operate.

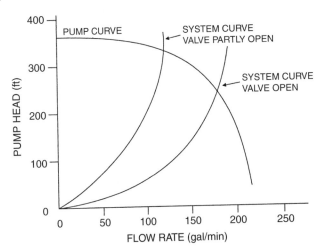

20. The pump that is depicted is going to be pumping in a system that will require its discharge head to be 200 feet. What will be the flow out of the pump?   _____

21. A valve is left partially shut in the flow line. How much has the flow decreased?   _____

22. The partially shut valve has changed the system. What will be the discharge pressure (discharge head) of the pump in the changed system?   _____

Note: Problems 23–25 deal with the airflow chart shown on the following page.

The airflow chart shown is a graph showing three different quantities. Knowing any two quantities related to the airflow will allow you to find a single point on the graph. You must be very careful to make sure that you are looking at the correct set of lines when you try to find a point on the graph, or read a value from the graph. The horizontal and vertical lines are related to the axes and their values are shown there. The values that the slanted lines represent are shown near an end of each line.

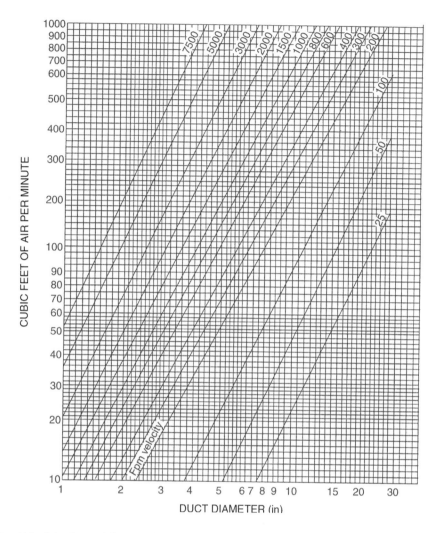

23. What is the minimum duct size needed to move 200 cu ft per minute and maintain duct velocity below 800 feet per minute? _____

24. A 4-inch diameter round duct has air moving at 600 feet per minute. What rate of airflow comes out of the duct? _____

25. In order to move 100 cu ft per minute down a 5-inch diameter duct, what velocity must the air have? _____

Bills

Unit 38 ESTIMATES AND BILLS

BASIC PRINCIPLES OF ESTIMATES AND BILLS

- Study and apply the principles of estimates and bills to the problems in this unit.

Customers often ask for itemized bills after work has been completed. At times they also want a written estimate for a possible job. It is important to be able to give these to the customer. The problems in this unit are designed to give practice in writing estimates and bills. For each problem, fill in the form and answer the related questions.

A bill is filled out for a number of reasons. One is to give you a record of what inventory was used. A bill also gives a record of how much time workers spent on various jobs. It also can be used by the customer to see what pieces were replaced.

To make it clear to everyone using the bill as a record, each bill should be clearly filled out. Make sure you write a clear description of the pieces used, the labor involved, or the basic charge made. Make sure that you put down the price per item when multiple items are used. This should go in the price column. The price per item times the number of items is the amount charged and should go in the amount column.

If the bill is an estimate, be sure to clearly mark that on the bill. It is also a good idea to put down how long the customer has to decide if he or she wants the job done for that price.

When filling out bills and estimates, it is important to fill out as much as possible and to be as clear as possible. You want someone else to read the sheet and understand it completely. Take your time to be certain that there is nothing that is uncertain or unclear.

PRACTICAL PROBLEMS

1. A customer asks for an estimate for the installation of a central air conditioner.
    A 3-ton unit is required. The estimate will show the estimated costs for parts
    and labor separately. The word *estimate* is to be written clearly on the form.

    A 3-ton unit $1,469.00
    Labor (15) hours . . . . $11.00 per hour

Keep Kool Refrigeration Co. 1000 Main St. Anytown, SC 29000 Customer:		Date Salesperson No. 3149		
QUANTITY	ITEM		PRICE	AMOUNT
ORIGINAL	Thank You			

a. If a 4% tax on parts is added, what is the additional charge?

a. _____

b. Two installers will install the unit. Each will work the same amount of time
    and will be paid $8.00 per hour. How much will each installer earn from
    this job?

b. _____

2. A refrigerator is repaired using the following:

   1 compressor $253.75
   1½ pounds refrigerant R-134a .... $1.70 per pound
   1 thermostat control $15.95
   Nuts, washers, grease $0.80
   5 hours labor .... $13.50 per hour

		Date			
Keep Kool Refrigeration Co. 1000 Main St. Anytown, SC 29000		Salesperson			
		No. 3149			
Customer:					

QUANTITY	ITEM		PRICE	AMOUNT
ORIGINAL	Thank You			

a. If the refrigerator owner does the work without hiring a technician, what will be the cost of the parts alone?

a. _____

b. The technician is paid $11.25 per hour. How much money does the company get for the labor on this job?

b. _____

c. How much does the technician make on this job?

c. _____

3. The installation of a 3½-ton central air-conditioning system needs the following:

     1 compressor, condenser, and evaporator $1,142.00

     1 concrete slab for condenser $32.00

     2½ pounds refrigerant R-134a charge . . . . $1.20 per pound

     2 foot drainpipe (1-inch diameter) . . . . $0.42 per foot

     30 foot connecting tubing . . . . $0.79 per foot

     1 plenum $47.50

     1 roll duct insulation $52.95

     1 roll duct tape $2.49

     1 thermostat $28.95

     35 foot thermostat wire $5.95

     Labor (3 installers—8 hours each) . . . . $12.75 per hour

Keep Kool Refrigeration Co.
1000 Main St.
Anytown, SC 29000

Customer:

Date

Salesperson

No.   3149

QUANTITY	ITEM	PRICE	AMOUNT

ORIGINAL   Thank You

a. It takes one installer 7 hours to insulate the ducts using the insulation and tape. How much would be saved if the ducts were already insulated?

a. _____

b. The company offers a 6% discount if paid in cash. How much is saved by paying cash?

b. _____

4. A technician makes an annual check on an oil burner. The charges for items and services are:

    1 gun nozzle $6.75
    1 air filter (14 inches × 20 inches) $1.90
    1 fuel filter cartridge $2.95
    Clean burner and make adjustments $15.00
    Check for water in fuel tank $4.25
    House call $25.00

Jones Heating Co. 100 Center St. Yourtown, PA 18000 Customer:			Date Salesperson No. 4285	
QUANTITY	ITEM		PRICE	AMOUNT
Customer's Copy	Thank You			

a. Seven dollars of the house call charge go for truck expenses. The remainder of the house call charge and the other costs (except parts) are for labor. If the technician takes 1½ hours to do the job, what is the labor cost per hour?

a. _____

b. If the homeowner changes the filters beforehand, what is the charge for the visit?

b. _____

5. An oil burner needs repairs. The technician replaces the parts listed:

      1 gun nozzle $6.65

      25 feet of $\frac{3}{8}$-inch copper fuel line .... $0.79 per foot

      1 fuel filter $6.95

      1 new oil pump $52.95

      1 pair oil gun electrodes .... $2.35 each

      Labor (2 hours) .... $9.50 per hour

Jones Heating Co. 100 Center St. Yourtown, PA 18000 Customer:		Date Salesperson No. 4285		
QUANTITY	ITEM		PRICE	AMOUNT
Customer's Copy	Thank You			

a. A rebuilt oil pump costs $29.00 and takes 1 hour to install. What is the bill if a rebuilt pump is used?

    a. _____

b. A 4% sales tax is charged for parts. How much does this add to the original bill?

    b. _____

6. A car air conditioner system is recharged. The cost is normally $69.95 but is currently $48.50. You may want the bill to show the regular price and the sale price so that the customer knows that it was a special. You also may want to note that this was a special in case a problem arises and you need to refund the full price—you will want to refund the sale price, and not the regular price. Include on the bill 5.5% sales tax. Write a bill for the recharging.

Jones Heating Co. 100 Center St. Yourtown, PA 18000 Customer:		Date Salesperson No. 4285	
QUANTITY	ITEM	PRICE	AMOUNT
Customer's Copy	Thank You		

7. A new motel with 15 rooms is under construction. The cost of each unit is $323.00. Each unit will require an electrical breaker at $8.25. An electrical wall socket, outlet box, and cover are needed for each unit at $3.97. Eighteen feet of electrical wire at $0.24 per foot are needed for each unit. The labor to install one unit is 4 hours, and the labor is charged at the rate of $31.00 per hour. There is a 6% sales tax on everything. Write a bill for the job of putting a through-the-wall unit in each room.

Keep Kool Refrigeration Co. 1000 Main St. Anytown, SC 29000 Customer:		Date Salesperson No. 3149		
QUANTITY	ITEM		PRICE	AMOUNT
ORIGINAL		Thank You		

8. You have a job to upgrade a unit on a refrigerated trailer to the new R-134a refrigerant. You set the rates as follows: reclaiming the old R-12 so it can be disposed, $17.00; flushing and replacing the seals on the system, $6.17; and recharging the new system, $6.25. Labor involved is 3 hours at $23.00 an hour. There is no sales tax for this bill. Write a bill for this job.

Keep Kool Refrigeration Co. 1000 Main St. Anytown, SC 29000 Customer:		Date Salesperson No. 3149	
QUANTITY	ITEM	PRICE	AMOUNT
ORIGINAL	Thank You		

9. You have been asked to convert a baseboard electric to a forced air heating system for a house. The charges for the customer are: removal of the old baseboard heaters, 3.5 hours at $19.00 per hour; cost of the ducts and plenum, $423.30; cost of the heat pump, $1,139.59; addition of a humidifier to the system, $325.00; diffusers and return duct grates, $62.50; hauling away trash, $125.00; and labor to install the system, 16 hours at $19.00 per hour. Write a bill for this conversion.

Jones Heating Co. 100 Center St. Yourtown, PA 18000 Customer:		Date Salesperson No. 4285	
QUANTITY	ITEM	PRICE	AMOUNT
Customer's Copy	Thank You		

10. To repair an oil burner system, a new nozzle at $8.75 is needed. The labor involved is 4 hours at $22.00 per hour. The 6% sales tax in the state is only on parts, not on labor. Write a bill for this repair job.

Jones Heating Co. 100 Center St. Yourtown, PA 18000 Customer:			Date Salesperson No. 4285	
QUANTITY	ITEM		PRICE	AMOUNT
Customer's Copy	Thank You			

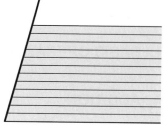

Appendix

Section I

DENOMINATE NUMBERS

Denominate numbers are numbers that include units of measurement. The units of measurement are arranged from the largest unit at the left to the smallest unit at the right.

Example: 6 yd 2 ft 4 in

All basic operations of arithmetic can be performed on denominate numbers.

I. EQUIVALENT MEASURES

Measurements that are equal can be expressed in different terms. For example, 12 in = 1 ft. If these equivalents are divided, the answer is 1.

$$\frac{1 \text{ ft}}{12 \text{ in}} = 1 \qquad \frac{12 \text{ in}}{1 \text{ ft}} = 1$$

To express one measurement as another equal measurement, multiply by the equivalent in the form of 1.

To express 6 inches in equivalent foot measurement, multiply 6 inches by 1 in the form of $\frac{1 \text{ ft}}{12 \text{ in}}$. In the numerator and denominator, divide by a common factor.

$$6 \text{ in} = \frac{\overset{1}{\cancel{6} \text{ in}}}{1} \times \frac{1 \text{ ft}}{\underset{2}{\cancel{12} \text{ in}}} = \frac{1}{2} \text{ ft or } 0.5 \text{ ft}$$

To express 4 feet in equivalent inch measurement, multiply 4 feet by 1 in the form of $\frac{12 \text{ in}}{1 \text{ ft}}$.

$$4 \text{ ft} = \frac{\overset{4}{\cancel{4} \text{ ft}}}{1} \times \frac{12 \text{ in}}{\underset{1}{\cancel{1} \text{ ft}}} = \frac{48 \text{ in}}{1} = 48 \text{ in}$$

Per means division, as with a fraction bar. For example, 50 miles per hour can be written $\frac{50 \text{ mi}}{1 \text{ hr}}$.

II. BASIC OPERATIONS

A. ADDITION

Example: 2 yd 1 ft 5 in + 1 ft 8 in + 5 yd 2 ft

1. Write the denominate numbers in a column with like units in the same column.

2. Add the denominate numbers in each column.

3. Express the answer using the largest possible units.

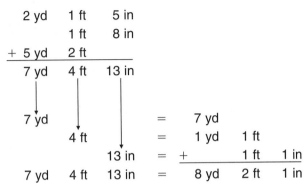

	2 yd	1 ft	5 in
		1 ft	8 in
+	5 yd	2 ft	
	7 yd	4 ft	13 in

7 yd = 7 yd
4 ft = 1 yd 1 ft
13 in = + 1 ft 1 in
7 yd 4 ft 13 in = 8 yd 2 ft 1 in

B. SUBTRACTION

Example: 4 yd 3 ft 5 in − 2 yd 1 ft 7 in

1. Write the denominate numbers in columns with like units in the same column.

	4 yd	3 ft	5 in
−	2 yd	1 ft	7 in

7 in is larger than 5 in

2. Starting at the right, examine each column to compare the numbers. If the bottom number is larger, exchange one unit from the column at the left for its equivalent. Combine like units.

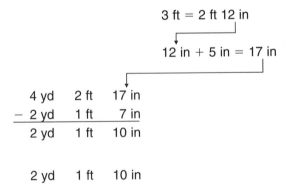

3 ft = 2 ft 12 in

12 in + 5 in = 17 in

3. Subtract the denominate numbers.

	4 yd	2 ft	17 in
−	2 yd	1 ft	7 in
	2 yd	1 ft	10 in

4. Express the answer using the largest possible units.

	2 yd	1 ft	10 in

C. MULTIPLICATION

—By a constant

Example: 1 hr 24 min × 3

1. Multiply the denominate number by the constant.

2. Express the answer using the largest possible units.

$$
\begin{array}{cc}
1\ \text{hr} & 24\ \text{min} \\
& \times\ 3 \\
\hline
3\ \text{hr} & 72\ \text{min}
\end{array}
$$

$$
\begin{array}{cc}
3\ \text{hr} & \\
& 72\ \text{min} \\
\hline
3\ \text{hr} & 72\ \text{min}
\end{array}
$$

= 3 hr
= 1 hr   12 min
= 4 hr   12 min

—By a denominate number expressing linear measurement

Example: 9 ft 6 in × 10 ft

1. Express all denominate numbers in the same unit.

$$9\ \text{ft}\ 6\ \text{in} = 9\frac{1}{2}\ \text{ft}$$

2. Multiply the denominate numbers. (This includes the units of measure, such as ft × ft = sq ft.)

$$9\frac{1}{2}\ \text{ft} \times 10\ \text{ft} =$$

$$\frac{19}{2}\ \text{ft} \times 10\ \text{ft} =$$

95 sq ft

—By a denominate number expressing square measurement

Example: 3 ft × 6 sq ft

1. Multiply the denominate numbers. (This includes the units of measure, such as ft × ft = sq ft and sq ft × ft = cu ft.)

3 ft × 6 sq ft = 18 cu ft

—By a denominate number expressing rate

Example: 50 mi per hr × 3 hours

1. Express the rate as a fraction using the fraction bar for *per.*

$$\frac{50\ \text{mi}}{1} \times \frac{3\ \text{hr}}{1}$$

2. Divide the numerator and denominator by any common factors, including units of measure.

$$\frac{50\ \text{mi}}{\cancel{1}} \times \frac{\cancel{3\ \text{hr}}^{\,3}}{1}$$

3. Multiply numerators.
    Multiply denominators.

$$\frac{150 \text{ mi}}{1} =$$

4. Express the answer in the remaining unit. 150 mi

D. DIVISION

—By a constant

Example: 8 gal 3 qt ÷ 5

1. Express all denominate numbers
    in the same unit.

    8 gal 3 qt = 35 qt

2. Divide the denominate number
    by the constant.

    35 qt ÷ 5 = 7 qt

3. Express the answer using the
    largest possible units.

    7 qt = 1 gal 3 qt

—By a denominate number expressing linear measurement

Example: 11 ft 4 in ÷ 8 in

1. Express all denominate numbers
    in the same unit.

    11 ft 4 in = 136 in

2. Divide the denominate numbers
    by a common factor. (This includes
    the units of measure, such as inches
    ÷ inches = 1.)

    136 in ÷ 8 in =
    $$\frac{\overset{17}{\cancel{136 \text{ in}}}}{\underset{1}{\cancel{8 \text{ in}}}} = \frac{17}{1} = 17$$

—By a linear measure with a square measurement as the dividend

Example: 20 sq ft ÷ 4 ft

1. Divide the denominate numbers.

    (This includes the units of measure,
    such as sq ft ÷ ft = ft.)

    20 sq ft ÷ 4 ft
    $$\frac{\overset{5 \text{ ft}}{\cancel{20 \text{ sq ft}}}}{\underset{1}{\cancel{4 \text{ ft}}}} = \frac{5 \text{ ft}}{1}$$

2. Express the answer in the remaining unit. 5 ft

—By denominate numbers used to find rate

Example: 200 mi ÷ 10 gal

1.   Divide the denominate numbers.

$$\frac{\overset{20\ mi}{\cancel{200\ mi}}}{\underset{1\ gal}{\cancel{10\ gal}}} = \frac{20\ mi}{1\ gal}$$

2.   Express the units with the fraction bar meaning *per*.

$$\frac{20\ mi}{1\ gal} = 20 \text{ miles per gallon}$$

Note: Alternate methods of performing the basic operations will produce the same results. The choice of method is determined by the individual.

Section II

EQUIVALENT VALUES

EQUIVALENT ENGLISH RELATIONSHIPS

LENGTH EQUIVALENTS

1/16 inch	=	0.0625	inch
1/8 inch	=	0.125	inch
3/16 inch	=	0.1875	inch
1/4 inch	=	0.25	inch
5/16 inch	=	0.3125	inch
3/8 inch	=	0.375	inch
7/16 inch	=	0.4375	inch
1/2 inch	=	0.5	inch
9/16 inch	=	0.5625	inch
5/8 inch	=	0.625	inch
11/16 inch	=	0.6875	inch
3/4 inch	=	0.75	inch
13/16 inch	=	0.8125	inch
7/8 inch	=	0.875	inch
15/16 inch	=	0.9375	inch
1 inch	=	0.0833	foot
2 inches	=	0.1666	foot
3 inches	=	0.25	foot
4 inches	=	0.3333	foot
5 inches	=	0.4166	foot
6 inches	=	0.5	foot
7 inches	=	0.5833	foot
8 inches	=	0.6666	foot
9 inches	=	0.75	foot
10 inches	=	0.8333	foot
11 inches	=	0.9166	foot

ENGLISH LENGTH MEASURE

1 foot (ft)	=	12 inches (in)
1 yard (yd)	=	3 feet (ft)
1 mile (mi)	=	1,760 yards (yd)
1 mile (mi)	=	5,280 feet (ft)

ENGLISH AREA MEASURE

1 square yard (sq yd)	=	9 square feet (sq ft)
1 square foot (sq ft)	=	144 square inches (sq in)
1 square mile (sq mi)	=	640 acres
1 acre	=	43,560 square feet (sq ft)

ENGLISH VOLUME MEASURE FOR SOLIDS

1 cubic yard (cu yd)	=	27 cubic feet (cu ft)
1 cubic foot (cu ft)	=	1,728 cubic inches (cu in)

ENGLISH VOLUME MEASURE FOR FLUIDS

1 quart (qt)	=	2 pints (pt)
1 gallon (gal)	=	4 quarts (qt)

ENGLISH VOLUME MEASURE EQUIVALENTS

1 gallon (gal)	=	0.133681 cubic foot (cu ft)
1 gallon (gal)	=	231 cubic inches (cu in)

ENGLISH WEIGHT (MASS) MEASURE EQUIVALENTS

1 pound (lb)	=	16 ounces (oz)

SI METRICS STYLE GUIDE

SI metrics is derived from the French name Le Système International d'Unités. The metric unit names are already in accepted practice. SI metrics attempts to standardize the names and usages so that students of metrics will have a universal knowledge of the application of terms, symbols, and units.

The English system of mathematics (used in the United States) has always had many units in its weights and measures tables that were not applied to everyday use. For example, the pole, perch, furlong, peck, and scruple are not used often. These measurements, however, are used to form other measurements, and it has been necessary to include the measurements in the tables. Including these measurements aids in the understanding of the orderly sequence of measurements greater or smaller than the less frequently used units.

The metric system also has units that are not used in everyday application. Only by learning the lesser-used units is it possible to understand the order of the metric system. SI metrics, however, places an emphasis on the most frequently used units.

In using the metric system and writing its symbols, certain guidelines are followed. For the student's reference, some of the guidelines are listed.

1. In using the symbols for metric units, the first letter is capitalized only if it is derived from the name of a person.

Example:

UNIT	SYMBOL	UNIT	SYMBOL
meter	m	Newton	N (named after Sir Isaac Newton)
gram	g	degree Celsius	°C (named after Anders Celsius)

Exceptions: The symbol for liter is L. This is used to distinguish it from the number one (1).

2. Prefixes are written with lowercase letters.

Example:

PREFIX	UNIT	SYMBOL
centi	meter	cm
milli	gram	mg

Exceptions:

PREFIX	UNIT	SYMBOL
tera	meter	Tm (used to distinguish it from the metric tonne, t)
giga	meter	Gm (used to distinguish it from gram, g)
mega	gram	Mg (used to distinguish it from milli, m)

3.   Periods are not used in the symbols. Symbols for units are the same in the singular and the plural (no "s" is added to indicate a plural).

Example: 1 mm   *not*   1 mm

3 mm   *not*   3 mms

4.   When referring to a unit of measurement, symbols are not used. The symbol is used only when a number is associated with it.

Example: The length of the room is   *not*   The length of the room is expressed in m.

expressed in meters.   (*The length of the room is 25 m* is correct.)

5.   When writing measurements that are less than one, a zero is written before the decimal point.

Example: 0.25 m   *not*   .25 m

6.   A space is left between the digits and the unit of measure.

Example: 5,179,232 mm   *not*   5,179,232mm

7.   Symbols for area measure and volume measure are written with exponents.

Example: 3 cm^2 *not*   3 sq cm   4 km^3 *not*   4 cu km

8.   Metric words with prefixes are accented on the first syllable. In particular, kilometer is pronounced "kill′-o-meter." This avoids confusion with words for measuring devices, which are generally accented on the second syllable, such as thermometer (ther-mom′-e-ter).

METRIC RELATIONSHIPS

The base units in SI metrics include the meter and the gram. Other units of measure are related to these units. The relationship between the units is based on powers of ten and uses these prefixes:

kilo (1,000)   centi (0.01)   milli (0.001)

These tables show the most frequently used units with an asterisk (*).

METRIC LENGTH MEASURE

10 millimeters (mm)*	= 1 centimeter (cm)*
100 centimeters (cm)	= 1 meter (m)*
1,000 meters (m)	= 1 kilometer (km)*

◆ To express a metric length unit as a smaller metric length unit, multiply by a positive power of ten such as 10; 100; 1,000; 10,000; and so on.

◆ To express a metric length unit as a larger metric length unit, multiply by a negative power of ten such as 0.1; 0.01; 0.001; 0.0001; and so on.

METRIC AREA MEASURE

100 square millimeters (mm^2)	= 1 square centimeter (cm^2)
10,000 square centimeters (cm^2)	= 1 square meter (m^2)
10,000 square meters (m^2)	= 1 square kilometer (km^2)

◆ To express a metric area unit as a smaller metric area unit, multiply by 100; 10,000; 1,000,000; and so on.

◆ To express a metric area unit as a larger metric area unit, multiply by 0.01; 0.0001; 0.000001; and so on.

METRIC VOLUME MEASURE FOR SOLIDS

1,000 cubic millimeters (mm^3)	= 1 cubic centimeter (cm^3)*
1,000,000 cubic centimeters (cm^3)	= 1 cubic meter (m^3)*
1,000,000,000 cubic meters (m^3)	= 1 cubic kilometer (km^3)

◆ To express a metric volume unit for solids as a smaller metric volume unit for solids, multiply by 1,000; 1,000,000; 1,000,000,000; and so on.

◆ To express a metric volume unit for solids as a larger metric volume unit for solids, multiply by 0.001; 0.000001; 0.000000001; and so on.

METRIC VOLUME MEASURE FOR FLUIDS

100 milliliters (mL)*	= 1 centiliter (cL)
100 centiliters (cL)	= 1 liter (L)*
1,000 liters (L)	= 1 kiloliter (kL)

◆ To express a metric volume unit for fluids as a smaller metric volume unit for fluids, multiply by 10; 100; 1,000; 10,000; and so on.

◆ To express a metric volume unit for fluids as a larger metric volume unit for fluids, multiply by 0.1; 0.01; 0.001; 0.0001; and so on.

METRIC VOLUME MEASURE EQUIVALENTS

1,000 cubic centimeters (cm^3)	= 1 liter (L)
1 cubic centimeter (cm^3)	= 1 milliliter (mL)

METRIC MASS MEASURE

10 milligrams (mg)*	= 1 centigram (cg)
100 centigrams (cg)	= 1 gram (g)*
1,000 grams (g)	= 1 kilogram (kg)*
1,000 kilograms (kg)	= 1 megagram (Mg)*

◆ To express a metric mass unit as a smaller metric mass unit, multiply by 10; 100; 1,000; 10,000; and so on.

◆ To express a metric mass unit as a larger metric mass unit, multiply by 0.1; 0.01; 0.001; 0.0001; and so on.

Metric measurements are expressed in decimal parts of a whole number. For example, one-half millimeter is written as 0.5 mm.

In calculating with the metric system, all measurements are expressed using the same prefixes. If answers are needed in millimeters, all parts of the problem should be expressed in millimeters before the final solution is attempted. Diagrams that have dimensions in different prefixes must first be expressed using the same unit.

ENGLISH–METRIC EQUIVALENTS
LENGTH MEASURE

1 inch (in)	=	25.4 millimeters (mm)
1 inch (in)	=	2.54 centimeters (cm)
1 foot (ft)	=	0.3048 meter (m)
1 yard (yd)	=	0.9144 meter (m)
1 mile (mi)	=	1.609 kilometers (km)
1 millimeter (mm)	=	0.03937 inch (in)
1 centimeter (cm)	=	0.3937 inch (in)
1 meter (m)	=	3.28084 feet (ft)
1 meter (m)	=	1.09361 yards (yd)
1 kilometer (km)	=	0.62137 mile (mi)

AREA MEASURE

1 square inch (sq in)	=	645.16 square millimeters (mm^2)
1 square inch (sq in)	=	6.4516 square centimeters (cm^2)
1 square foot (sq ft)	=	0.092903 square meter (m^2)
1 square yard (sq yd)	=	0.836127 square meter (m^2)
1 square millimeter (mm^2)	=	0.00155 square inch (sq in)
1 square centimeter (cm^2)	=	0.155 square inch (sq in)
1 square meter (m^2)	=	10.76391 square feet (sq ft)
1 square meter (m^2)	=	1.19599 square yards (sq yd)

VOLUME MEASURE FOR SOLIDS

1 cubic inch (cu in)	=	16.387064 cubic centimeters (cm^3)
1 cubic foot (cu ft)	=	0.028317 cubic meter (m^3)
1 cubic foot (cu ft)	=	28,317 cubic centimeters (m^3)
1 cubic yard (cu yd)	=	0.764555 cubic meter (m^3)
1 cubic centimeter (cm^3)	=	0.061024 cubic inch (cu in)
1 cubic meter (m^3)	=	61,023 cubic inches (cu in)
1 cubic meter (m^3)	=	35.314667 cubic feet (cu ft)
1 cubic meter (m^3)	=	1.307951 cubic yards (cu yd)

VOLUME MEASURE FOR FLUIDS

1 gallon (gal)	=	3,785.411 cubic centimeters (cm^3)
1 gallon (gal)	=	3.785411 liters (L)
1 quart (qt)	=	0.946353 liter (L)
1 ounce (oz)	=	29.57353 cubic centimeters (cm^3)
1 cubic centimeter (cm^3)	=	0.000264 gallon (gal)
1 liter (L)	=	0.264172 gallon (gal)
1 liter (L)	=	1.056688 quarts (qt)
1 cubic centimeter (cm^3)	=	0.033814 ounce (oz)

MASS MEASURE

1 pound (lb)	=	0.453592 kilogram (kg)
1 pound (lb)	=	453.59237 grams (g)
1 ounce (oz)	=	28.349523 grams (g)
1 ounce (oz)	=	0.02835 kilogram (kg)
1 kilogram (kg)	=	2.204623 pounds (lb)
1 gram (g)	=	0.002205 pound (lb)
1 kilogram (kg)	=	35.273962 ounces (oz)
1 gram (g)	=	0.035274 ounce (oz)

Section III

INSTRUMENTS FOR MEASURING

At times, very accurate measurements are needed. Two devices used to get accurate measurements are the vernier caliper and the micrometer. The challenge is to read scales. So, let us make sure that we can use and read the scales.

MEASURING WITH A VERNIER CALIPER

To read the scale on the vernier caliper, look first at where the 0 line is lining up on the main scale. This gives the first two numbers. Then look along the vernier scale to find the line that most closely lines up with one of the lines on the main scale. This will give the next two digits.

Example: Determine the reading on the vernier caliper in the photo. The reading is needed in inches, so the lower two scales are used. (The upper two scales are used for millimeter readings.)

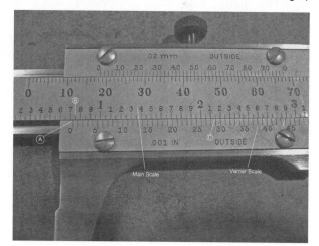

A. The 0 line on the vernier scale is just past the 7, which is *before* the large 1 on the main scale. Since it is before the large 1, the reading begins 0.70.

B. The 0 line is *before* the short mark indicating 7.5, so the next part of the reading will be less than 50.

C. Looking down the vernier scale, we see that the 29 line on the vernier exactly matches a line on the main scale (in this case, the 2.15 line). At the bottom of the vernier is an indication .001 in. This means that each mark on the vernier scale represents 0.001 inch. Since the 29 line matches, this represents 0.029 in.

D. Combine the readings from step A and step C.

$$0.7 + 0.029 = 0.729$$

The final reading is 0.729 inch.

Note: If the 0 line had been past the little mark in part B, the reading would have been 0.75 plus 0.029 to give a final reading of 0.779 inch.

MEASURING WITH A MICROMETER

The micrometer is another device used to make accurate measurements. To read the scale, the first part of the reading is determined by what is uncovered by the rotating barrel as it unwinds. The rest of the reading is determined by where the main scale line lines up with the markings on the barrel.

The numbers on the main (stationary) part of the scale give the first digit and part of the rest of the reading. The printed numbers on the main scale of this micrometer are tenths of an inch. Notice that the distance between two numbers is divided into 4 quarters. Each mark (division) on the main scale between the tenths marks represents 0.025 inch. The rotating barrel of the micrometer has 25 equal divisions on it. One complete revolution uncovers one division on the main scale. One revolution covers 0.025 inch, so each mark on the barrel represents 0.001 inch.

Example: Determine the reading on the micrometer in the photo.

A.   The barrel (rotating part) has uncovered the 2. This represents 0.2.

B.   The barrel has also uncovered two divisions past the 2. This means that the reading is going to be 2 × 0.025 or 0.05 more than the 0.2.

C.   The main scale line lines up with the 6 on the rotating barrel. This gives an additional 6 × 0.001 or 0.006 inch.

D.   Combine these three parts to give the total reading.

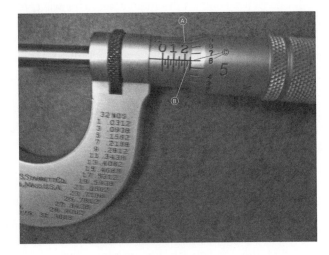

0.2 + 0.05 + 0.006 = 0.256 inch.

If the rotating barrel in our picture had uncovered 3 divisions after the 2, the reading would be 0.2 + 0.075 + 0.006 or 0.281 inch. If the rotating barrel in this picture had shown a number such as 14, then the reading would be 0.2 + 0.05 + 0.014 giving a total of 0.264 inches. As can be seen, the micrometer has to be read carefully. Some micrometers have lines on the back side of the main part of the scale. This set of lines is a vernier that will add an additional digit if the marks on the barrel do not line up exactly with the main scale line. The additional digit is determined by a line on the vernier lining up with a line on the rotating barrel.

Section IV

FORMULAS

PERIMETER

Square
$P = 4s$

P = perimeter
s = side

Rectangle
$P = 2l + 2w$

P = perimeter
l = length
w = width

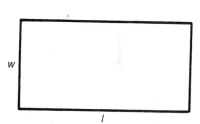

Triangle
$P = a + b + c$

P = perimeter
a = first side
b = second side
c = third side

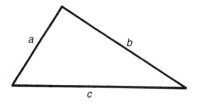

Circle
$C = 2\pi r$
$C = \pi D$

$r = \dfrac{D}{2}$

C = circumference
π = 3.1416
r = radius
D = diameter

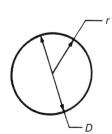

AREA

Square

$A = s^2$

A = area
s = side

Rectangle

$A = lw$

A = area
l = length
w = width

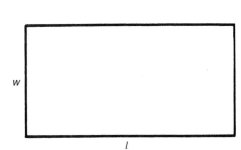

Triangle

$A = \frac{1}{2}bh$

A = area
b = base
h = height

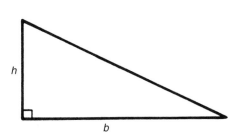

Trapezoid

$A = \frac{1}{2}(b_1 + b_2)h$

A = area
b_1 = first base
b_2 = second base
h = height

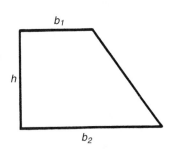

Circle

$A = \pi r^2$

A = area
π = 3.1416
r = radius
D = diameter

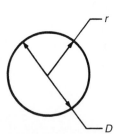

VOLUME

Rectangular Solid

$V = lwh$

V = volume
l = length
w = width
h = height

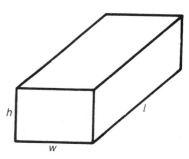

Cylindrical Solid

$V = \pi r^2 h$

V = volume
π = 3.1416
r = radius
D = diameter
h = height

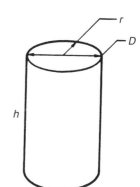

TEMPERATURE

$°C = 5/9\ (°F - 32)$ $°C$ = degrees Celsius
$°F = 9/5\ (°C) + 32$ $°F$ = degrees Fahrenheit
$K = °C + 273$ K = Kelvins
$°R = °F + 460$ $°R$ = degrees Rankine

ELECTRICAL

Ohm's Law

$E = IR$

E = voltage (in volts)
I = current (in amperes)
R = resistance (in ohms)

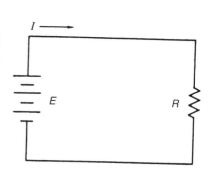

Power Formula

$P = IE$

P = power (in watts)
I = current (in amperes)
E = voltage (in volts)

Resistance in Series

$$R = R_1 + R_2 + \ldots R$$ = resistance (in ohms)

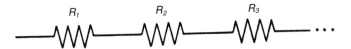

Resistance in Parallel

$$R = \frac{1}{\dfrac{1}{R_1} + \dfrac{1}{R_2} + \ldots}$$ R = resistance (in ohms)

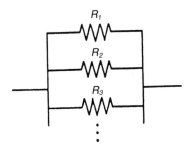

Capacitance in Series

$$C = \frac{1}{\dfrac{1}{C_1} + \dfrac{1}{C_2} + \ldots}$$ C = capacitance (in microfarads)

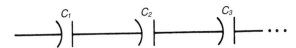

Capacitance in Parallel

$$C = C_1 + C_2 + \ldots$$ C = capacitance (in microfarads)

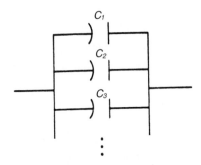

GAS LAW

General Gas Law

$$\frac{P_1 V_1}{T_1} = \frac{P_2 V_2}{T_2}$$

P = pressure (absolute)
V = volume
T = temperature (absolute)

Boyle's Law

$$P_1 V_1 = P_2 V_2$$

P = pressure (absolute)
V = volume

Charles' Law

$$\frac{P_1}{T_1} = \frac{P_2}{T_2}$$

P = pressure (absolute)
T = temperature (absolute)

Guy-Lussac's Law

$$\frac{V_1}{T_1} = \frac{V_2}{T_2}$$

V = volume
T = temperature (absolute)

Section V

HEAT TRANSFER MULTIPLIERS

Note: The heat transfer multiplier is found by multiplying the U factor (the amount of heat transferred through 1 sq ft of structure for each degree temperature difference between the inside and outside surfaces) by the design temperature difference. The units for the heat transfer multiplier are British thermal units per hour per square foot, so all areas must be in units of square feet.

TYPE OF STRUCTURE	DESIGN TEMPERATURE DIFFERENCE		
	25°F	70°F	75°F
Walls — wood frame with sheathing and siding or other veneer			
3½ inches insulation (R-11)	3.5	5	5
3½ inches insulation + 1 inch polystyrene sheathing (R-16)	3.1	3.5	3.8
6 inches insulation (R-19)	2.6	2.8	2.9
Brick with 3½ inches insulation	3.4	4.5	4.9
Brick with 3½ inches insulation + 1 inch polystyrene sheathing	3.0	3.4	3.7
Metal siding with 1½ inches polystyrene	3.3	4.3	4.7
Metal siding with 4 inches polystyrene	1.3	3.3	3.6
Ceiling — under vented roof			
3½ inches insulation (R-11)	2.5	6	6
6 inches insulation (R-19)	1.5	4	4
9½ inches insulation (R-30)	1.0	2.2	2.4
Floor			
Slab on ground	No heat loss	No heat loss	No heat loss
No insulation over crawl space/basement	5	16	17
6 inches insulation over crawl space/basement	1	3.2	3.4
Windows			
Single pane	35	105	110
Double pane	25	70	75
Single pane + storm window	25	60	65
Double pane (fixed)	25	60	65
Double pane + storm window	12	35	38
Doors			
Insulated core, weather-stripped	5.3	81	86
Sliding glass door, double glass	25	90	95

Section VI

TRIGONOMETRIC FUNCTIONS

TRIGONOMETRIC FUNCTIONS

Angle	Sine	Cosine	Tangent	Angle	Sine	Cosine	Tangent
1°	0.0175	0.9998	0.0175	46°	0.7193	0.6947	1.0355
2°	0.0349	0.9994	0.0349	47°	0.7314	0.6820	1.0724
3°	0.0523	0.9986	0.0524	48°	0.7431	0.6691	1.1106
4°	0.0698	0.9976	0.0699	49°	0.7547	0.6561	1.1504
5°	0.0872	0.9962	0.0875	50°	0.7660	0.6428	1.1918
6°	0.1045	0.9945	0.1051	51°	0.7771	0.6293	1.2349
7°	0.1219	0.9925	0.1228	52°	0.7880	0.6157	1.2799
8°	0.1392	0.9903	0.1405	53°	0.7986	0.6018	1.3270
9°	0.1564	0.9877	0.1584	54°	0.8090	0.5878	1.3764
10°	0.1736	0.9848	0.1763	55°	0.8192	0.5736	1.4281
11°	0.1908	0.9816	0.1944	56°	0.8290	0.5592	1.4826
12°	0.2079	0.9781	0.2126	57°	0.8387	0.5446	1.5399
13°	0.2250	0.9744	0.2309	58°	0.8480	0.5299	1.6003
14°	0.2419	0.9703	0.2493	59°	0.8572	0.5150	1.6643
15°	0.2588	0.9659	0.2679	60°	0.8660	0.5000	1.7321
16°	0.2756	0.9613	0.2867	61°	0.8746	0.4848	1.8040
17°	0.2924	0.9563	0.3057	62°	0.8829	0.4695	1.8807
18°	0.3090	0.9511	0.3249	63°	0.8910	0.4540	1.9626
19°	0.3256	0.9455	0.3443	64°	0.8988	0.4384	2.0503
20°	0.3420	0.9397	0.3640	65°	0.9063	0.4226	2.1445
21°	0.3584	0.9336	0.3839	66°	0.9135	0.4067	2.2460
22°	0.3746	0.9272	0.4040	67°	0.9205	0.3907	2.3559
23°	0.3907	0.9205	0.4245	68°	0.9272	0.3746	2.4751
24°	0.4067	0.9135	0.4452	69°	0.9336	0.3584	2.6051
25°	0.4226	0.9063	0.4663	70°	0.9397	0.3420	2.7475
26°	0.4384	0.8988	0.4877	71°	0.9455	0.3256	2.9042
27°	0.4540	0.8910	0.5095	72°	0.9511	0.3090	3.0777
28°	0.4695	0.8829	0.5317	73°	0.9563	0.2924	3.2709
29°	0.4848	0.8746	0.5543	74°	0.9613	0.2756	3.4874
30°	0.5000	0.8660	0.5774	75°	0.9659	0.2588	3.7321
31°	0.5150	0.8572	0.6009	76°	0.9703	0.2419	4.0108
32°	0.5299	0.8480	0.6249	77°	0.9744	0.2250	4.3315
33°	0.5446	0.8387	0.6494	78°	0.9781	0.2079	4.7046
34°	0.5592	0.8290	0.6745	79°	0.9816	0.1908	5.1446
35°	0.5736	0.8192	0.7002	80°	0.9848	0.1736	5.6713
36°	0.5878	0.8090	0.7265	81°	0.9877	0.1564	6.3138
37°	0.6018	0.7986	0.7536	82°	0.9903	0.1392	7.1154
38°	0.6157	0.7880	0.7813	83°	0.9925	0.1219	8.1443
39°	0.6293	0.7771	0.8098	84°	0.9945	0.1045	9.5144
40°	0.6428	0.7660	0.8391	85°	0.9962	0.0872	11.4301
41°	0.6561	0.7547	0.8693	86°	0.9976	0.0698	14.3007
42°	0.6691	0.7431	0.9004	87°	0.9986	0.0523	19.0811
43°	0.6820	0.7314	0.9325	88°	0.9994	0.0349	28.6363
44°	0.6947	0.7193	0.9657	89°	0.9998	0.0175	57.2900
45°	0.7071	0.7071	1.0000	90°	1.0000	0.0000	

Glossary

Absolute pressure—The total pressure consisting of the gauge pressure plus the atmospheric pressure.

Absolute temperature—The temperature measured on an absolute scale that uses zero as the temperature at which all molecular motion stops.

Ampere—The unit of electric current.

Atmospheric pressure—The pressure exerted by the gases in the atmosphere upon the earth.

Bending radius—The radius of the circle formed by the arc made when tubing is bent.

British thermal unit—A unit for a quantity of heat. It is the quantity of heat needed to raise the temperature of one pound of water one degree Fahrenheit.

Capacitor—An electrical device used to store electric charges.

Combustion chamber—The part of a furnace where the fuel is burned.

Compression ratio—The ratio of the pressure of the gas leaving a compressor to the pressure of the gas going into the compressor.

Compressor—A device used in refrigerators that takes the refrigerant gas at low pressure and compresses or squeezes it into a gas at high pressure.

Condenser—The part of the refrigerator that takes the hot compressed gas from the compressor and allows it to cool to a liquid.

Condensing pressure—The pressure for a given temperature at which the gas becomes a liquid.

Conduit—A tube used to carry and protect electrical wires.

Connecting rod—The part of a compressor that connects the piston with the crankshaft.

Cooling load—The amount of heat that must be removed from a space in a given amount of time.

Damper—A valve for controlling airflow.

Density—A measure of the compactness of a substance. It is the mass of the substance per unit volume.

Diffuser—A device for deflecting or spreading out the airflow.

Duct—A tube or channel through which air is moved.

Elbow—A section of a pipe or duct that is bent at an angle.

Electrodes—Parts of an oil burner gun. Two pieces of metal set so there is a gap between them. An electric spark jumps across the gap and ignites the oil.

Energy Efficiency Rating (EER)—A measure comparing the energy used by a device with the work gotten from that device. For air-conditioning units, it is defined as the ratio of the heat removed in Btu/hr divided by the energy used in watts.

Evaporator—The part of a refrigerator where the liquid refrigerant vaporizes. Heat is absorbed as the liquid changes to a gas.

Felt wiper—A piece of felt that is in contact with a rotating shaft. The felt is impregnated with oil to lubricate the turning shaft.

Filter-drier—A device for removing small particles and water from the refrigerant.

Fin comb—A device that looks like a comb and is used to straighten the bent metal fins on condensers or evaporators.

Flaring cone—A device used for expanding the end of tubing when making connections.

Flue—A passageway used to carry away smoke and burned gases from a furnace.

Gasket—A device or material used to form a seal between two objects when they are joined.

Gauge pressure—The pressure measured by and read from a gauge. It is the difference between the inside pressure and atmospheric pressure.

Ground source heat pump—A heat pump that uses the ground (or a body of water) as the heat source in the heating mode or the heat sink (place to put the heat) in the cooling mode.

Heat load—The amount of heat that must be added to a space in a given amount of time.

Heat loss—The amount of heat that flows through a boundary from a heated space to an unheated space.

Heat pump—A reversible system that can heat or cool a space.

High pressure side—The portion of a refrigeration system where the refrigerant is under high pressure. This portion extends from the exhaust of the compressor through the condenser to the expansion valve.

Humidifier—A device that puts moisture into the air.

Infrared heater—A device giving off heat that is hot enough to be sensed by the skin, but is not hot enough to glow.

Latent heat of fusion—The amount of heat energy given off as a liquid becomes a solid with no change in temperature or pressure. This is also the amount of heat energy absorbed as the solid becomes a liquid.

Latent heat of vaporization—The amount of heat energy absorbed as a liquid becomes a gas with no change in temperature or pressure. This is also the amount of heat energy given off as the gas becomes a liquid.

Low pressure side—The portion of a refrigeration system where the refrigerant is at low pressure. This portion extends from the expansion valve through the evaporator to the intake of the compressor.

Microfarad—The unit used to measure the electric capacitance.

Motor bearing—A support for the rotating part of a motor.

Mullion heater—The electrical heating element mounted in the part of the refrigerator between the two doors (mullion). The heater is used to prevent sweating.

Parallel circuit—An electric circuit connected in such a manner that the current flows through at least two paths.

Piston—The part of a compressor or motor that moves up and down inside the cylinder.

Piston ring—A thin, narrow ring that fits in a groove around the piston and makes a very tight fit inside the cylinder.

Plenum—A chamber for moving air. A large duct.

Pressure gauge—A device for measuring the pressure of a gas by comparing it with atmospheric pressure.

PVC tubing—Plastic tubing made of polyvinylchloride.

R value—A measure of the resistance to heat flow through a material.

Radiant heater—A device that emits heat energy from the surface in the form of waves without the movement of air or water.

Reed valve—A valve used in a compressor. The valve is a flat piece of metal that is placed over the opening. One end of the metal strip is fastened and the other end is free to move up or down to open or close the opening.

Refrigerant—The substance used in a refrigerating system that changes from a gas to a liquid and back again and in doing so transfers heat energy.

Relay capacitor—A relay is a switching device that uses a small electrical signal to activate the switch. The capacitor is the storage device that causes the relay to activate when it becomes totally charged.

Resistance heater—A type of heater that produces heat by allowing a current to flow through a special wire.

Rotary compressor—A device that compresses gases using a rotating motion rather than a reciprocating (up and down) motion.

Rotor—The part that spins in a motor or rotary compressor.

Running capacitor—The storage device that provides a properly timed current to a motor while it is running. The current must be timed properly to cause the motor to turn.

Saturated vapor pressure—The pressure for a given temperature at which a refrigerant can exist as both a vapor and a liquid.

Series circuit—An electric circuit connected in such a manner that the current must flow in a single path.

Shim—A piece of uniform size material such as a washer or a varying thickness material such as a wedge used for filling space.

Smoke pipe—A thin-walled sheet metal tubing used in chimneys.

Spreader—A diffuser that increases the width of an airflow.

Stator—The stationary part of an electric motor.

Strap—A strip of metal used to support a suspended pipe or duct.

Stretchout—The pattern for a duct or piece of metal. The pattern is used to cut a flat piece of sheet metal that is then bent to form a duct.

Tee—A joint in pipes or ducts in the shape of the letter T.

Ton refrigeration unit—A unit measuring the amount of refrigeration. One ton of refrigeration will remove in 24 hours the amount of heat needed to melt one ton of ice.

U factor—A measure of the amount of heat flow through a barrier.

V-belt—A belt that often connects motors and fans. The belt is named for the V-shaped contact surface.

Vaporizing pressure—A pressure for a given temperature at which a liquid can become a gas.

Volt—The unit of electric potential or electromotive force.

Watt—The unit of electric power.

Weather stripping—A strip of material used to seal joints that do not fit tightly together such as doors or windows.

Y—A joint in pipes or ducts in the shape of the letter Y.

ANSWERS TO ODD-NUMBERED PROBLEMS

SECTION 1   WHOLE NUMBERS

UNIT 1   ADDITION OF WHOLE NUMBERS

1. 976
3. 8,979
5. 990 meters
7. 879
9. 7,298

11. 11,745 feet
13. 46 feet
15. 29 feet
17. 25 feet

19. 9,056 cu ft
21. 46 feet
23. 555 pounds
25. 36,442 miles

UNIT 2   SUBTRACTION OF WHOLE NUMBERS

1. 64
3. 711
5. 1,269 quarts
7. 331
9. 715

11. 589 sq cm
13. 40°F
15. 18 pounds
17. 124 cu ft/min

19. 127 feet
21. 86 gallons
23. 921 sq ft
25. 40 cu ft/min

UNIT 3   MULTIPLICATION OF WHOLE NUMBERS

1. 219
3. 467,372
5. 17,232 millimeters
7. 1,235
9. 46,828

11. 483,426 sq ft
13. 1,008 connectors
15. $272.00
17. 4,932 pounds

19. 962,000 Btu
21. 380 feet
23. 210 feet
25. 1,932 man-hours

UNIT 4   DIVISION OF WHOLE NUMBERS

1. 23
3. 26
5. 356 Btu
7. 43
9. 67

11. 36 pounds
13. 83 pounds
15. 475 pounds
17. 45 rolls

19. 13 days
21. 14 sheets
23. 725 cu ft
25. 23 air conditioners

UNIT 5   COMBINED OPERATIONS WITH WHOLE NUMBERS

1. 1,873
3. 425
5. 3,348,660
7. 24
9. 5,937 inches

11. 27,848,634 seconds
13. 1,622 cu ft
15. 76 valves
17. a. 12 minutes
    b. 5 times

19. 215 cu ft/min
21. 38,000 Btu/hr
23. 25 ducts
25. a. 288,000 Btu
    b. 12,000 Btu

SECTION 2   COMMON FRACTIONS

UNIT 6   ADDITION OF COMMON FRACTIONS

1. $^8/_{13}$
3. $^3/_8$
5. $2^1/_5$
7. $1^{12}/_{25}$
9. $45^{11}/_{16}$

11. $1^{31}/_{32}$ inches
13. $18^5/_8$ inches
15. $6^3/_8$ inches
17. $22^1/_{16}$ inches

19. $140^3/_4$ inches
21. $31^2/_3$ feet
23. $17^1/_4$ hours
25. $27\ ^{13}/_{16}$ inches

UNIT 7   SUBTRACTION OF COMMON FRACTIONS

1. $^3/_7$
3. $3^3/_5$
5. $4^{27}/_{35}$
7. $^{38}/_{75}$
9. $^{21}/_{40}$

11. $^3/_{16}$ inch
13. $2^1/_4$ inches
15. $4^5/_8$ feet
17. $21^5/_8$ inches

19. $3^3/_8$ inches
21. $2^1/_8$ inches
23. $45^3/_4$ inches
25. $^{13}/_{16}$ inches

UNIT 8   MULTIPLICATION OF COMMON FRACTIONS

1. $^3/_{16}$
3. $^8/_{15}$
5. $2^4/_5$
7. $1^4/_5$
9. $^7/_{18}$ foot
11. $2^3/_8$ inches

13. $61^1/_5$ inches
15. $5,580
17. $21^1/_2$ inches
19. $256^1/_2$ inches
21. 44¢

23. a. $4^1/_2$ hours
    b. 3 hours
    c. $^1/_2$ hour
    d. 1 hour
25. $^{11}/_{320}$ pound

UNIT 9   DIVISION OF COMMON FRACTIONS

1. $^{18}/_{35}$
3. $1^4/_5$
5. $^{49}/_{78}$
7. $^1/_{16}$ inch
9. $1^5/_7$ weeks

11. $27^1/_9$ Btu/hr
13. 14 systems
15. 4 strips
17. 8 washers

19. 3 joists
21. $4^1/_2$ days
23. 27 dehumidifiers
25. 7 segments

UNIT 10   COMBINED OPERATIONS WITH COMMON FRACTIONS

1. $5^{34}/_{45}$
3. $2^{1}/_{3}$
5. $^{5}/_{24}$
7. $^{14}/_{15}$
9. $^{13}/_{24}$ yard

11. $^{3}/_{4}$ pound
13. $50^{1}/_{8}$ inches
15. $^{5}/_{16}$ inches
17. $2^{5}/_{8}$ inches
19. $4^{1}/_{4}$ inches

21. $78^{3}/_{4}$ inches
23. $11^{5}/_{8}$ inches
25. a. $6^{3}/_{4}$ hours
    b. $33^{3}/_{4}$ hours
    c. $11^{23}/_{27}$ days

SECTION 3   DECIMAL FRACTIONS

UNIT 11   ADDITION OF DECIMAL FRACTIONS

1. 574.63
3. 393.633
5. 1,249.4 cu in
7. 145.32
9. 1,758.0763

11. 825.152 pounds
13. 169.8 lb/sq in
15. 0.625 inch
17. 70.3 Btu/lb

19. 0.25 inch
21. 3.18
23. 62.8 pounds
25. 12.72 amperes

UNIT 12   SUBTRACTION OF DECIMAL FRACTIONS

1. 123.42
3. 1.294
5. 42.767 sq in
7. 89.8
9. 618.015

11. 4.2573 m^2
13. 0.143 lb/cu ft
15. 1.358 lb/cu ft
17. 0.143 inch

19. 0.0186 lb/sq in
21. 0.294 pound
23. 34.7 pounds
25. 727.41 psi

UNIT 13   MULTIPLICATION OF DECIMAL FRACTIONS

1. 2,251.092
3. 5.486112
5. 7.568739 pounds
7. 3.5070637
9. 0.011084

11. 40.9307 cm^2
13. 74.4 inches
15. 3.858 lb/sq in
17. 14.62 pounds

19. 42.6875 Btu
21. 847.88 Btu
23. 13.3114 amperes
25. 1,315.02 Btu

UNIT 14   DIVISION OF DECIMAL FRACTIONS

1. 3.52
3. 0.418
5. 7.522 gallons
7. 531.5
9. 3.082

11. 33.35 inches
13. 1.835 amperes
15. 3 tons
17. 1.65 gallons/hr

19. 0.2849 (rounded off)
21. 70.7 Btu
23. 0.097 inch
25. 2,116.8 cu ft

UNIT 15   DECIMAL AND COMMON FRACTION EQUIVALENTS

1. 0.1875
3. 0.875
5. 0.175
7. 0.2667
9. 0.1667
11. 8.4135

13. 439.74
15. 467.749
17. 1.102 inches
19. No.
   0.0305 inch is less than
   $\frac{1}{32}$ inch

21. 1.379 inches
23. $46.71
25. 3.32 gallons

UNIT 16   COMBINED OPERATIONS WITH DECIMAL FRACTIONS

1. 20.8861
3. 0.0436345
5. 90.2088 centimeters
7. 7.893 feet
9. 19.1622 meters

11. 2.5395 cu in (rounded off)
13. 0.375
15. 1.61 inches
17. 21,483 Btu

19. 0.015625 inch clearance
21. 58.5 Btu
23. 7,450.114 cu ft/hr
25. $253.21

SECTION 4   RATIO AND PROPORTION

UNIT 17   RATIO

1. 3/5
3. 5/3
5. 3/7
7. 4/3
9. 22/7

11. 7/18
13. 1/2
15. 9/7
17. 3/4

19. 17/20
21. 3.5/1
23. 3/32
25. 5/6

UNIT 18   PROPORTION

1. 15
3. 6.8
5. 27
7. 11.7
9. 1,597.5 pounds

11. 82.8 pounds
13. 56.3
15. 7.2 feet
17. 12,960 cu in

19. 0.021 inch of water
21. 850 rpm
23. 7.5 tons
25. 186.7 pounds (rounded off)

SECTION 5   PERCENT, PERCENTAGE, AND DISCOUNT

UNIT 19   PERCENT AND PERCENTAGE

1. 0.072
3. 0.837
5. 4.15
7. 143%
9. 160
11. $456.40

13. 6 hours
15. $24,000.00
17. 0.36 pound
19. $300.00
21. 33%

23. a. $2,340.00
    b. $210.60
    c. $153.04
    d. $2,703.64
25. $0.73 a foot

UNIT 20   DISCOUNTS

1. $706.80
3. $46.73
5. $506.25
7. $1,344.64
9. a. $3,515.50
    b. $3,752.50

11. $208.18
13. a. $606.83
    b. $634.73
15. $2,932.16
17. $4.74
19. $1,373.67

21. $9.59
23. $55.43
25. a. $1,068.00
    b. $1,164.00
    c. $96.00

SECTION 6   DIRECT MEASURE

UNIT 21   EQUIVALENT UNITS OF TEMPERATURE MEASURE

1. 25°C
3. 77.8°C
5. 143.6°F
7. 393K
9. 545°R
11. 537°R

13. 5°C
15. 10°C
17. 20°C
19. a. 18.3°C
    b. 60°C

21. a. 77°F
    b. 41°F
23. 4.4°C
25. a. 82.2°C
    b. 75.6°C

UNIT 22   ANGULAR MEASURE

1. 35°
3. 128°
5. 90°
7. 120°
9. a. 45°
    b. 45°

11. a. 50°
    b. 50°
13. a. 60°
    b. 60°
    c. 60°
15. 227°

17. 90°
19. a. 90°
    b. 119°
21. 38°
23. 75°
25. 150°

UNIT 23   UNITS OF LENGTH MEASURE

1. 60 inches
3. 86 inches
5. 175 centimeters
7. 7 centimeters
9. 5 feet
11. $4\frac{1}{6}$ feet

13. 4.65 meters
15. 5.5 meters
17. 31 inches
19. 45 inches

21. a. $12\frac{1}{2}$ feet $\times$ $12\frac{3}{4}$ feet
    b. 10 feet $\times$ $10\frac{1}{3}$ feet
    c. $10\frac{1}{6}$ feet $\times$ $13\frac{1}{2}$ feet
    d. $9\frac{1}{6}$ feet $\times$ $14\frac{1}{4}$ feet
23. 31 inches
25. 1,046 centimeters

UNIT 24   EQUIVALENT UNITS OF LENGTH MEASURE

1. 22.86 centimeters
3. 48.26 centimeters
5. 2.13 meters
7. 5.906 inches
9. 2 feet 9.858 inches
11. 22 feet 11.59 inches

13. 15.24 centimeters
15. a. 9 feet 10.1 inches
    b. 16 feet 4.9 inches
17. 9.8425 inches
19. 11.6472 meters left

21. a. 121.92 centimeters
    b. 60.96 centimeters
23. a. 50.8 centimeters
    b. 40.64 centimeters
25. 4.72 fins/centimeter

UNIT 25   LENGTH MEASURE

1. 19 feet 10 inches
3. 16 feet $10\frac{1}{2}$ inches
5. a. 1,040 centimeters
   b. 10.4 meters
7. 26 feet $6\frac{3}{4}$ inches

9. a. 1.23 meters
   b. 0.67 meter
11. a. 6.55 centimeters
    b. 8.1 centimeters
13. 9 feet 9 inches
15. 48 inches

17. 56 inches
19. 30.27 inches
21. $50\frac{3}{4}$ inches
23. 2.513 inches
25. 0.193 inch

SECTION 7   COMPUTED MEASURE

UNIT 26   AREA MEASURE

1. 64 sq in
3. a. $\frac{9}{16}$ sq in
   b. $25\frac{5}{16}$ sq in
5. 80 ft^2
7. $245\frac{5}{8}$ sq ft

9. 6.16 m^2
11. 38.485 in^2
13. 0.64 in^2
15. 60 sq ft
17. 234 sq ft

19. 34.54 m^2
21. circular
23. 12 inches
25. 60 sq in

UNIT 27 EQUIVALENT UNITS OF AREA MEASURE

1. 31 sq yd
3. 12.5 sq ft
5. a. 5,760 sq in
   b. 40 sq ft
7. 2.6896 m²
9. 227 sq ft (rounded off)

11. 1.77 sq in
13. 0.0346361 m²
15. 89.7579 sq ft
17. a. 19,932.812 cm²
    b. 1.9932812 m²

19. 8,300 sq in
21. 464.52 cm²
23. 248 sq in
25. 12.9 sq ft

UNIT 28 RECTANGULAR VOLUMES

1. 720 cu ft
3. 1,232 cu ft
5. 0.3 m³
7. 1,900.4 cu ft
9. 2,962 cu yd

11. a. 6,415.5 cu in
    b. 3.71 cu ft
13. 2,807.5 cu ft
15. 27.91 cu ft
17. 155 cu ft/min

19. 0.75 foot
21. 1,493 cu yd/hr
23. 66.75 cu ft
25. 1,079 rolls

UNIT 29 CYLINDRICAL VOLUMES

1. 2,463.014 cu in
3. 1,809.562 cu ft
5. 502.656 cm³
7. 49.088 cu in
9. 8.639 cu in

11. 955.255 cu in
13. 439,824 cm³
15. a. 2,770.891 cu in
    b. 1.604 cu ft
17. 18.653 cu ft

19. a. 61,072.704 cu in
    b. 264.384 gallons
21. 4.937 gallons
23. 174.533 cu ft
25. 243,698.589 cm³

SECTION 8 FORMULAS

UNIT 30 OHM'S LAW AND ELECTRICAL RELATIONSHIPS

1. 2 ohms
3. 2.5 amperes
5. 2.67 amperes
7. 9.58 ohms
9. ¾ or 0.75 ampere

11. 2,990 watts
13. 227.27 volts
15. 0.5 ampere
17. 115.08 volts

19. 360 ohms
21. 388 ohms
23. 16 microfarads
25. No

UNIT 31 GAS LAWS AND TEMPERATURE OF MIXTURES

1. 12 cu in
3. 7.5 kPa
5. 23.33 psia
7. 11.38 psia
9. 620°F

11. 81°C
13. 0.28 cu in
15. 1,667.1 kPa; No
17. 59.5°C

19. 57.6 psia
21. 20.17°F
23. 0.11 cu in
25. 121.25°F

UNIT 32 HEAT LOAD CALCULATIONS

1. 15,240 Btu/hr

3. 1,754.9 Btu/hr

5. 4,286 Btu/hr

7. 5,580.0 Btu/hr

9. 2,530.5 Btu/hr

11. 643.7 Btu/hr

13. 3,031.7 Btu/hr

15. 22,444 Btu/hr

17. 3,349 Btu/hr

19. 10,239.7 Btu/hr

21. 48,435.8 Btu/hr

23. 36,715.2 Btu/hr

25. 30,348 Btu/hr

SECTION 9 STRETCHOUTS AND LENGTHS OF ARCS

UNIT 33 STRETCHOUTS OF SQUARE AND RECTANGULAR DUCTS

1. a. 48 inches
   b. 18 inches
3. a. 32 inches
   b. $26\frac{3}{4}$ inches
5. a. 36 inches
   b. 36 inches
7. a. $35\frac{3}{4}$ inches
   b. $25\frac{1}{2}$ inches
9. a. 90.8 centimeters
   b. 75 centimeters

11. a. $108\frac{1}{4}$ inches
    b. $24\frac{3}{4}$ inches
13. a. $37\frac{1}{4}$ inches
    b. 36 inches
15. a. 100 centimeters
    b. 43 centimeters
17. a. 90.25 centimeters
    b. 100 centimeters
19. a. $36\frac{3}{4}$ inches
    b. $32\frac{1}{2}$ inches

21. a. $96\frac{3}{4}$ inches
    b. 23 inches
23. a. $44\frac{3}{4}$ inches
    b. 32 inches
25. a. 117.9 centimeters
    b. 86 centimeters

UNIT 34 STRETCHOUTS OF CIRCULAR DUCTS

1. a. $18\frac{7}{8}$ inches
   b. 24 inches
3. a. 75.4 centimeters
   b. 50 centimeters
5. a. 102.42 centimeters
   b. 110 centimeters
7. a. 64.4 centimeters
   b. 75 centimeters
9. a. $25\frac{3}{8}$ inches
   b. 36 inches

11. a. $31\frac{13}{16}$ inches
    b. 30 inches
13. a. 40 inches
    b. 36 inches
15. a. $35\frac{3}{16}$ inches
    b. $30\frac{7}{8}$ inches
17. a. $24\frac{11}{16}$ inches
    b. 28 inches
19. a. 71.52 centimeters
    b. 70 centimeters

21. a. 66.46 centimeters
    b. 90 centimeters
23. a. 27 inches
    b. 36 inches
25. a. 69.57 centimeters
    b. 120 centimeters

UNIT 35 LENGTHS OF ARCS OF CIRCLES

1. 1.57 inches
3. 15.71 feet
5. 150°
7. 4.19 centimeters
9. 15.708 inches
11. 21°
13. 3.14 inches

15. a. $3\frac{15}{16}$ inches
   b. $11\frac{3}{4}$ inches
17. a. $3\frac{9}{16}$ inches
   b. $10\frac{5}{8}$ inches
19. a. 54.978 centimeters
   b. 14.766 centimeters

21. a. 15.708 centimeters
   b. 34.558 centimeters
23. a. 9.739 centimeters
   b. 25.84 centimeters
25. a. 4.32 inches
   b. 10.799 inches

SECTION 10   TRIGONOMETRY

UNIT 36   TRIGONOMETRIC FUNCTIONS

1. 39°
3. 28°
5. 22°
7. a. 72°
   b. 144°
9. 7.99 centimeters

11. 8.08 feet
13. 7.71 feet
15. 31°
17. 5.22 inches
19. 5.37 inches

21. 58 feet
23. a. 14°
   b. 12.4 feet
25. length 5.456 feet
   height 5.8512 feet

SECTION 11   GRAPHS

UNIT 37   GRAPHS AND GRAPHING

1. July
3. 3 technicians
5. a. 6 technicians
   b. 1 technician

7.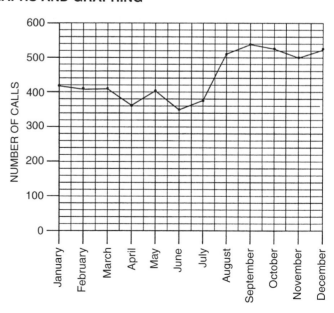

9. 4 part-time technicians
11. 33.5 Btu
13. 15 Btu
15. Overcharged,
    take some out

17. a. No
    b. Yes
    c. Yes
19. 48%

21. 60 gpm decrease
23. 5.6-inches diameter
    or larger
25. 600 ft/min

SECTION 12 BILLS

UNIT 38 ESTIMATES AND BILLS

1. a. $58.76
   b. $60.00
3. a. $144.69
   b. $98.72

5. a. $74.20
   b. $3.55

7. $7,370.29
9. $2,445.89